Management and Industrial Engineering

Series Editor

J. Paulo Davim, Department of Mechanical Engineering, University of Aveiro, Aveiro, Portugal

This series fosters information exchange and discussion on management and industrial engineering and related aspects, namely global management, organizational development and change, strategic management, lean production, performance management, production management, quality engineering, maintenance management, productivity improvement, materials management, human resource management, workforce behavior, innovation and change, technological and organizational flexibility, self-directed work teams, knowledge management, organizational learning, learning organizations, entrepreneurship, sustainable management, etc. The series provides discussion and the exchange of information on principles, strategies, models, techniques, methodologies and applications of management and industrial engineering in the field of the different types of organizational activities. It aims to communicate the latest developments and thinking in what concerns the latest research activity relating to new organizational challenges and changes world-wide. Contributions to this book series are welcome on all subjects related with management and industrial engineering. To submit a proposal or request further information, please contact Professor J. Paulo Davim, Book Series Editor, pdavim@ua.pt

More information about this series at http://www.springer.com/series/11690

Aravind Raj Sakthivel · Jayakrishna Kandasamy ·
J. Paulo Davim
Editors

Managing Supply Chain Risk and Disruptions: Post COVID-19

Springer

Editors
Aravind Raj Sakthivel
Department of Manufacturing Engineering
Vellore Institute of Technology
Vellore, Tamil Nadu, India

Jayakrishna Kandasamy
Department of Manufacturing Engineering
Vellore Institute of Technology
Vellore, Tamil Nadu, India

J. Paulo Davim
Department of Mechanical Engineering
University of Aveiro
Aveiro, Portugal

ISSN 2365-0532 ISSN 2365-0540 (electronic)
Management and Industrial Engineering
ISBN 978-3-030-72574-7 ISBN 978-3-030-72575-4 (eBook)
https://doi.org/10.1007/978-3-030-72575-4

This Springer imprint is published by the registered company Springer Nature Switzerland AG
The registered company address is: Gewerbestrasse 11, 6330 Cham, Switzerland

We dedicate this book to all front line workers, researchers, and practitioners who involved in action against COVID-19 for their dedicated service and sacrifices they made during this pandemic period

Thank You.

Foreword

I am glad to write a preface for the highly relevant book on the trending topic *Managing Supply Chain Risk and Disruptions: Post COVID-19* edited by Aravind Raj, Jayakrishna, and Paulo. The pandemic has taught us several lessons and the book chapters attempted to compile the list of challenges in several sectors including retail, agriculture, pharmaceutical, and humanitarian. As the title indicates the book unfolds a contingency plan, long-term mitigation mechanisms with the support of technologies are useful leads for further thoughts.

The Chap. 1 listed the challenges in the global supply chains using the SCOR framework in the volatility, uncertainty, complexity, and ambiguous scenarios. Specific review considering black swan events including previous disruptions are a good addition to a reader in understanding the importance of the topic as well as engage with the author's views on the use of practices with the support of data and technology.

The next four chapters on preparedness and assessment of COVID-19 on different sectors are good evidence for readers to sense the contextual perspectives through primary sources discussed by the authors. It is interesting to see the role of disintermediation, local autonomy, and use of information technology as well as future research directions mentioned by the authors. In a similar vein, the authors articulated well food insecurity issues and overall loss that occurred due to pandemic with the data from the Indian agriculture sector. Logistical issues due to whiplash effects are argued well within the humanitarian supply chain along with the list of skillsets needed to deal with future disasters.

The next four chapters outlay the dynamics of the supply chain and the author's perspective on the agile practices from the end to end supply chain, adaptability, decision making, resilient supply network, and emergent action plans. The use of basic multi-criteria decision-making techniques was wisely used by the authors to support stakeholders engaged in the relief operations. On the network side, building resilient networks is the trendy term used globally, and using the Indian case as an example the authors spelled out the basic barriers in manufacturing, operations, and distribution in the global supply and distribution network. Post COVID-19 action plans in terms of capacity enhancement have also been proposed by the authors.

The final two chapters narrate the useful applications of technologies in the healthcare supply chain and generally argue the potential pathways to enhance robustness in the Industry 4.0 scenario.

Overall I find the book will be a useful reference to both researchers and practitioners to enhance their application and contextual knowledge in the topic of risk and disruptions as well as encourage future researchers to contribute to the scholarly debates on building resilient supply networks.

Nachiappan Subramanian
Professor of Operations and Logistics
Management and Supply Chains
University of Sussex Business School
Brighton, UK

Preface

Supply chain (SC) risks are complicated and can be classified into operational and disruption risks. Pandemic outbursts are an unusual case of supply chain (SC) risks causing huge disruption with its existence, propagation, and high level of uncertainty. In recent days, the COVID-19 outbreak represents to be one of the major disruptions encountered during the past decades which is "breaking many global supply chains" severely (Araz et al. 2020). This book summarizes the effect of COVID-19 on the global supply chain by comprehending the solutions to these questions.

- What were the challenges faced by supply chain providers during COVID-19?
- What supply chain strategies and practices have contributed to the COVID-19 crisis?
- What can supply chain professionals do to begin preparing for a future in which this type of disruption may happen again?

This book also suggests the resilient strategies that can be adopted by supply chain organizations, for quick recovery and reestablishing their networks. Further, this book focuses on the challenges faced and how supply chain networks managed to be agile, during the current global pandemic COVID-19. The role of digital technologies, such as machine learning, artificial intelligence, big data, blockchain, and other IIoT tools in building resilient supply chains will also be highlighted.

Impact in supply chain, challenges faced during this pandemic, issues in agricultural sector, supply chain adoption strategies, decision support system to cater emergency solution, supply chain post COVID-19, and new technologies for health care supply chain were addressed in this book

This book will be a useful resource to researchers, students, and supply chain professionals by offering solutions to restructure the global supply chains post COVID-19.

Vellore, India — Aravind Raj Sakthivel
Vellore, India — Jayakrishna Kandasamy
Aveiro, Portugal — J. Paulo Davim

Acknowledgements

The editors would like to acknowledge the help of all the people involved in this project and, more specifically, to the authors and reviewers that took part in the review process. Without their support, this book would not have become a reality.

First, the editors would like to thank each one of the authors for their contributions. Our sincere gratitude goes to the chapter's authors who contributed their time and expertise to this book.

Second, the editors wish to acknowledge the valuable contributions of the reviewers regarding the improvement of quality, coherence, and content presentation of chapters.

Aravind Raj Sakthivel
School of Mechanical Engineering
Vellore Institute of Technology
Vellore, Tamil Nadu, India

Jayakrishna Kandasamy
School of Mechanical Engineering
Vellore Institute of Technology
Vellore, Tamil Nadu, India

J. Paulo Davim
Department of Mechanical Engineering
University of Aveiro
Aveiro, Portugal

Contents

Editors and Contributors

About the Editors

Dr. Aravind Raj Sakthivel is an Associate Professor in the School of Mechanical Engineering at the Vellore Institute of Technology University, India. He received his doctorate degree in Industrial Engineering from the National Institute of Technology, Tiruchirappalli in 2014. His research area is focused on Agile manufacturing process and their applications in manufacturing industries. Recent work focuses on Industrial Systems Engineering, Additive Manufacturing, Industry 4.0 projects. He has published 20 International journals papers in leading SCI/ SCOPUS Indexed journals and 34 papers in reputed conference proceedings; he is co-author for four book chapters and edited two book. He is reviewer of reputed international journals. He has received several awards including Doctoral Scholarship and Best Paper Award.

Dr. Jayakrishna Kandasamy is an Associate professor in the School of Mechanical Engineering at the Vellore Institute of Technology University, India. Dr. Jayakrishna's research is focused on the design and management of manufacturing systems and supply chains to enhance efficiency, productivity and sustainability performance. More recent research is in the area of developing tools and techniques to enable value creation through sustainable manufacturing, including methods to facilitate more sustainable product design for closed-loop material flow in industrial symbiotic setup. He has published 50 journal publications in leading SCI/SCOPUS Indexed journals, 22 book chapters, and 85 refereed conference proceedings, 2 books in CRC/Springer Series. Dr. Jayakrishna's team has received numerous awards in recognition for the quality of the work that has been produced. He teaches undergraduate and graduate courses in the manufacturing and industrial systems area and his initiatives to improve teaching effectiveness have been recognized through national awards. He has also been awarded Institution of Engineers (India)-Young Engineer Award in 2019, awarded as Best Researcher for 2016–2019 consecutively by VIT University Vellore, and Distinguished Researcher Award in the field of Sustainable Systems

Engineering in 2019 by International Institute of Organized Research. He is the Coordinator of Circular Economy Club, VIT University, and serving as Academic Editor in the journal of Mathematical Problems in Engineering, Hindawi Publications.

Dr. J. Paulo Davim is a Full Professor at the University of Aveiro, Portugal. He is also distinguished as honorary professor in several universities/colleges/institutes in China, India and Spain. He received his Ph.D. degree in Mechanical Engineering in 1997, M.Sc. degree in Mechanical Engineering (materials and manufacturing processes) in 1991, Mechanical Engineering degree (5 years) in 1986 from the University of Porto (FEUP), the Aggregate title (Full Habilitation) from the University of Coimbra in 2005 and the D.Sc. (Higher Doctorate) from London Metropolitan University in 2013. He is a Senior Chartered Engineer in the Portuguese Institution of Engineers with an MBA and Specialist titles in Engineering and Industrial Management as well as in Metrology. He is also Eur Ing by FEANI, Brussels, and Fellow of IET (FIET), London. He has more than 30 years of teaching and research experience in Manufacturing, Materials, Mechanical and Industrial Engineering, with special emphasis on Machining and Tribology. He also has interest in Management, Engineering Education and Higher Education for Sustainability. He has guided large numbers of postdoc, Ph.D. and master's students as well as coordinated and participated in several financed research projects. He has received several scientific awards and honours. He has worked as evaluator of projects for European Research Council (ERC) and other international research agencies as well as examiner of Ph.D. thesis for many universities in different countries. He is the Editor-in-Chief of several international journals, Guest Editor of journals, books Editor, book Series Editor and Scientific Advisory for many international journals and conferences. Presently, he is an Editorial Board member of 30 international journals and acts as reviewer for more than 100 prestigious Web of Science journals. In addition, he has also published as editor (and co-editor) more than 200 books and as author (and co-author) more than 15 books, 100 book chapters and 500 articles in journals and conferences (more than 280 articles in journals indexed in Web of Science core collection/h-index 60+/11500+ citations, SCOPUS/h-index 64+/14000+ citations, Google Scholar/h-index 82+/24000+ citations). He has been listed in World's Top 2% Scientists by Stanford University study.

Contributors

Rohit Agrawal Department of Production Engineering, National Institute of Technology, Tiruchirappalli, Tamil Nadu, India

Md. Zahir Uddin Arif Faculty of Business Studies, Department of Marketing, Jagannath University, Dhaka, Bangladesh

Manimuthu Arunmozhi Energy Research Institute, Nanyang Technological University, Singapore, Singapore

Ravi Shankar Bhakat Management Development Institute Murshidabad (MDIM), Murshidabad, India

Krishna Kumar Dadsena IBS, Hyderabad, Telangana, India

Manavalan Ethirajan School of Mechanical Engineering, Vellore Institute of Technology, Vellore, Tamil Nadu, India

Selvakumar Gurusamy Mechanical Engineering, SSN College of Engineering, Chennai, Tamil Nadu, India

P. Kalpana Department of Mechanical Engineering, Indian Institute of Information Technology Design & Manufacturing (IIITDM), Kancheepuram, Tamil Nadu, India

V. Kamala College of Engineering, Department of Industrial Engineering, Anna University, Chennai, Tamil Nadu, India

Jayakrishna Kandasamy Mechanical Engineering, VIT University, Chennai, Tamil Nadu, India;
School of Mechanical Engineering, Vellore Institute of Technology, Vellore, Tamil Nadu, India

R. Kiran Kumar Volvo India Private Limited, Bangalore, India

K. Lakshmi College of Engineering, Department of Industrial Engineering, Anna University, Chennai, Tamil Nadu, India

Nitish Maan Department of Mechanical Engineering, National Institute of Technology Warangal, Warangal, India

Siva Kumar Mahalingam Mechanical Engineering, Vel Tech Rangarajan Dr. Sagunthala R&D Institute of Science and Technology, Chennai, Tamil Nadu, India

Vijaya Kumar Manupati Department of Mechanical Engineering, National Institute of Technology Warangal, Warangal, India

Deepak Mathivathanan Centre for Logistics and Supply Chain Management, Loyola Institute of Business Administration, Chennai, Tamil Nadu, India

K. Mathiyazhagan Operations & Quantitative Method, Thiagarajar School of Management, Madurai, Tamilnadu, India

Biswajita Mohanty Vellore Institute of Technology Vellore, Vellore, Tamil Nadu, India

Lenin Nagarajan Mechanical Engineering, Vel Tech Rangarajan Dr. Sagunthala R&D Institute of Science and Technology, Chennai, Tamil Nadu, India

Maciel M. Queiroz Postgraduate Program in Business Administration, Paulista University–UNIP, São Paulo, Brazil

Kotha Raj Kumar Reddy Supply Chain Analytics - SME, Chennai, Tamil Nadu, India

K. Sivakumar Department of Mechanical Engineering, Siddharth Institute of Engineering and Technology (SIETK), Puttur, Andhra Pradesh, India

B. A. Srinivasa MEI College, Bangalore, India

Atour Taghipour Faculty of International Business, Normandy University, Normandy, France

R. Vezhavendan Department of Manufacturing Engineering, School of Mechanical Engineering, Vellore Institute of Technology, Vellore, Tamil Nadu, India

S. Vinodh Department of Production Engineering, National Institute of Technology, Tiruchirappalli, Tamil Nadu, India

Abbreviations

3PL	Third-Party Logistics
AHP	Analytic Hierarchy Process
AI	Artificial Intelligence
COVID-19	Corona Virus Disease 2019
DEMATEL	Decision-Making Trial and Evaluation Laboratory
DSS	Decision Support System
FAO	Food and Agriculture Organization
FMCG	Fast Moving Consumer Goods
GSCM	Global supply-chain management
HCSC	Health Care Supply Chain
HSC	Humanitarian supply chain
IFRC	International Federation of Red Cross
INR	Indian Rupee
IT	Information Technology
MCDM	Multiple-Criteria Decision-Making
MI	Machine Learning
NGO	Non-Governmental Organizations
PCAR	Percentage Change in Arrival
PCP	Percentage Change in Price
PROMEETHE	Preference Ranking Organization Method for Enrichment Evaluation
QFD	Quality Function Deployment
SARS-CoV-2	Severe Acute Respiratory Syndrome Corona Virus 2
SC	Supply Chain
SWOT	Strengths, Weaknesses, Opportunities, and Threats
TOPSIS	Technique for Order of Preference by Similarity to Ideal Solution
VIKOR	VIseKriterijumska Optimizacija I Kompromisno Resenje
VUCA	Volatility, Uncertainty, Complexity, and Ambiguity

Chapter 1
Impact of COVID-19 on Global Supply Chain Management

Manimuthu Arunmozhi, R. Kiran Kumar, and B. A. Srinivasa

Abstract The outbreak of the COVID-19 pandemic is an exceptional shock to the global supply chain management. With the continuation of nation-wide lockdown, worldwide economic downfall, and interruption of demand and supply chains, the economy is likely to face an extended period of downturn. Since this outbreak caused long-term impact in the global supply chain, we have to enquire and solve the short-term impact to overcome the crisis in the current scenario. Moreover, the world is becoming more VUCA in nature, and pandemic is one of the white swan events which created huge impact in our daily life. This chapter discusses the risk factors, disruptions, and changes in the strategies in global supply practices. Further, companies should be prepared for the sudden hazard, disaster, health crisis, and emergency conditions for the steady running of the supply chain. The organizations must be proactive and face the challenge by planning and introducing new methods and technologies. Also, in a positive note, the outbreak of COVID-19 will help us to adopt and invent new strategies to be faced in future without breaking the supply chain face the black swan events soon.

Keywords COVID-19 · Pandemic · Global supply chain · VUCA · Black swan

1.1 Introduction

As the world is becoming more connected it gives opportunities to develop the business to new horizon in the VUCA world. The demand for products, goods, and services increases steadily. However, all of these were brought to a grinding halt

M. Arunmozhi (✉)
Energy Research Institute, Nanyang Technological University, Singapore, Singapore
e-mail: arunmozhi.m@ntu.edu.sg

R. Kiran Kumar
Volvo India Private Limited, Bangalore, India

B. A. Srinivasa
MEI College, Bangalore, India

A. R. Sakthivel et al. (eds.), *Managing Supply Chain Risk and Disruptions: Post COVID-19*, Management and Industrial Engineering,
https://doi.org/10.1007/978-3-030-72575-4_1

by the COVID-19 outbreak. Due to COVID-19 outbreak nation-wide there were lockdowns, airspace and borders were closed, witnessing an uncommon disturbance to the global economy, irrespective of size and level of supply chain operations globally (Araz et al. 2020). Further, COVID-19 continues to spread across the globe, causing disruption and uncertainty to business operations. Despite the exceptional steps and increasing efforts made by governments, the consequence of COVID-19 outbreak is being noticed strongly with every single day and stalking the businesses growth (Alghababsheh and Gallear 2020). According to Ivanov and Dolgui (2020), 75% of companies have experienced supply chain disruptions because of COVID-19 limiting the logistics activities significantly. Further, the volatility in the supply chain practices; uncertainty in the decision-making; complexity in co-ordination; and ambiguity in information sharing makes global supply chain a perfect candidate for VUCA (Bennett and Lemoine 2014).

Generally, suppliers are located across the globe. In this hard situation, the business companies facing the epidemic outbreaks have a certain interrogation, like how long the supply chain will undergo this disruption and a supply chain to recover after an epidemic outbreak (Ivanov and Dolgui 2020). For example, accepting the timely shortages, using pandemic plan of actions, and modifying the operations will help to overcome supply chain disruption. However, no one anticipated the severity of COVID-19 outbreak which led to high disruption and uncertainty in the global supply chain practices. Therefore, in this chapter, the study discusses about the risk factors, impacts, strategies, and challenges in future regarding the impact of COVID-19 in global supply chain (Ali et al. 2020). Global supply chain management (GSCM) is the scattering of stocks and services, by maximizing the profit and minimizing the waste in all transnational companies (Koberg and Longoni 2019). The difference between GSCM and SCM is that GSCM concentrates on transnational business corporation, while other concentrates on national companies. The GSCM focuses on the four distinct areas: operations management, supply management, logistics, and marketing. These are classified according to the area of concentration (Manuj et al. 2014). Further, global supply chains are systems that can extend across many continents and countries for sourcing and providing stocks and services. The nature of a GSCM is generally accessing low expense country and mentioning the achievement of product and services from countries with abundant human resources and economic production costs (Choi et al. 2020). Also, GSCM involves company, purchaser, supplier tiers, and the suppliers from different countries across the global. Due to COVID-19, the global supply chain faces many risks and disruptions in the field of market, logistics, and supply (Tang 2006). The raw materials from the suppliers play a major role in the supply chain. Around the world, most of the raw materials suppliers are from China; it is largely affected by the pandemic and the industries are shut for a very long period (Deloitte 2020). Therefore, the impact of COVID-19 on global supply chain was substantially significant and many countries are vulnerable to the indirect effects of the COVID-19 outbreak (Guan et al. 2020).

1.2 Global Supply Chains Composition

In businesses, we must ensure two important things in supply chain: one, the supply chain should be achievable, and the other is results should be delivered on time. The supply chain risks differ with the nature of the business, size of the business, and the products involved (Srinivas and Sreedharan 2018). The process flow of supply chain management starts with the product or service and ends at the delivery of the product and it is the utilization by the end customer. In our everyday lives, we use and consume million things. While delivering these products, the supply chain management creates a tuneful and efficient environment (Liu et al. 2016). If any break in this chain happens, it causes disruption in the system. The supply chain management is made of few components which are important and critical to the system. The explanations of the above-mentioned components are discussed in brief.

1.2.1 Planning

Planning is the basic step in every process. It is important to finalize the strategies and set up in their places before the beginning of the entire supply chain. Check the need for the product, and monitor the cost, feasibility, manpower, and profit, etc., which are important. Without proper planning, it is impossible to achieve long-term and effective benefits. We must spend enough time for this phase. Once the plans are finalized, consider the advantages and disadvantages to proceed further. Every business needs a process flowchart or plan to design the model (Peidro et al. 2009). Planning helps to find out the need and supply trends in the global market, and this in turn helps to lead an efficient GSCM (Sumner et al. 2020).

1.2.2 Information

The social networking is dominating the today's world by series flow of information. Updating the current trends is one of the essential that a business stays side by side with the latest information in all aspects of its production (Flint 2004). The information should be properly understood and timely disseminated by knowing the current market supply trends and the need for the product (Musa et al. 2014). Information plays a crucial role in knowledge-based economy and if we ignore any aspects of information in business it will bring down the prospects of the business (Rejikumar et al. 2019).

1.2.3 Source

Raw materials from the suppliers play a very important role in GSCM systems. The products which are sold to the end user are made with the help of different set of raw materials. The raw materials should be cost-effective with required quality. So, the supplier needs to deliver the raw materials on time within the proposed budget. If the supplier is unable to deliver, the businesses suffer and obtain negative honor (Anderson et al. 2007). The good quality resource from the supplier gives good quality products so that the supplier can maintain good reputation in the market. Thus, the supplier plays a strong role in supply chain management system.

1.2.4 Inventory

Inventory management is one of the essential parts in supply chain. It includes all the resources and raw materials needed for the production (Sreedharan et al. 2018). Inventory should be regularly monitored to update the stock. If the inventory is not properly maintained, it will affect the function of the supply chain and goods cannot be delivered on time (Peidro et al. 2009; Liu et al. 2016).

1.2.5 Production

Production is one of the essential aspects in GSCM. It is possible only when all plans are executed, and other components cope up with other. Production is followed with series of final preparation like test and packing before delivering the final products (Raghuram et al. 2020).

1.2.6 Location

All business needs a location which is profitable and flourish the business as well. A business cannot survive if the location do not have demand for the raw materials. Hence location should be connected to the essential resources for business' production prosperity. Availability of manpower is also considered while starting up a new business venture (Liu et al. 2016).

1.2.7 *Transportation*

Transportation is the prominent term in supply chain from manufacturing to distributing the semi-finished/finished product to the market. Further, on time delivery of goods is mandatory to sustain smooth business process (Mentzer et al. 2004). Also, it should ensure zero damage and no loss in transit. The company must work toward secured transit process while delivering the goods. The businesses which pay great attention to transportation management will benefit from the production and customer (Jia and Hu 2011).

1.2.8 *Return of Goods*

Facilities for the return of faulty goods with response to the consumer grievance also create a strong supply chain along with the other components. Machine may also become faulty once in a lifetime because no one is infallible. Customers expect return of goods under certain circumstances. Even the best quality machines have unavoidable control lapses. By following customer complaints, as strong business process they have to recall the product and issue an apology. This maintains goodwill and create good customer bonding. If any break and weak lines are identified, it should be dealt with immediately. Management as well as teams at various levels must put great emphasis on all the components (Hasani et al. 2012). For effective planning of execution, the business needs to be studied carefully with respect to demand and supply. Supply chain management system is very critical system to ensure consumer requirements and timely manufacture of goods.

1.3 VUCA World and Black Swan Events

The supply chain that acts in timely and flexible way will have the agility and resilience to withstand the turbulent times of today's VUCA world (Packowski 2014), where VUCA stands for volatility, uncertainty, complexity, and ambiguity.

Volatility: Change in dynamics and its nature; change in speed of forces and catalysts and its nature.

Uncertainty: The absence of anticipation, the hope for surprise, and the sense of attention and grasping of issues and events.

Complexity: A greater number of forces, problem my tying, no domino effect, and uncertainty flowing in the organizations.

Ambiguity: Reality out of focus, likely misinterpretations, and the domino effect confusion.

The VUCA term is gaining more fame as the rapid changes in global trade and commerce from ground zero to supply chain plans in every day. Volatility is both the speed of change and the frequency of change. Changes can be instigated by your own company, your competitors, or by a natural phenomenon, such as a pandemic or earthquake. Uncertainty is lack of predictability, which makes it impossible to define requirements (Bennett and Lemoine 2014). When you cannot define what you need to succeed, it is then impossible to optimize operations. Difficulty is the multitude of forces and confusing issues and the blurring of roles. Ambiguity is haziness—the inability to get a good read on a situation because it is in flux. There are two types of disruption upsetting the world; they are innovation and crisis. Innovation disruption is the normal cycle of new product and service development, while crisis disruption is the kind of "black swan" event that we are facing with the COVID-19 pandemic. Nobody saw it coming and it has "massive transformational impact". The pandemic is forming companies into four types, based on how they were doing in December 2019 to now; they are:

Success to success: These are companies that were doing well in December and now are doing even better. Retail stores is one example—business is booming, and its stock is up now over the end of 2019.

Success to failure: These companies were looking forward to a good 2020 but have had the rug pulled out from under them. Airline carriers and hospitality industry players are in this group.

Failure to success: Flat in December, companies in this group have experienced an unexpected surge in business due to the pandemic. Many sanitation companies were now able to meet the demand for its disinfection products, wipes. Some experts say the demand became five times the normal volume.

Failure to quit: Struggling already in December, these companies are now calling it quits. The key is to be able to regenerate and be flexible. Optionality is achieved by creating numerous scenarios and being able to act crucial between them as circumstances read out.

VUCA creates uncertainty, which means predicting the future is impossible. In a normal supply chain network, design enterprise managers would look at four- to five-year requirements, capital expenditures, and so on, and adjust the operation to meet the expected demands. In other words, designing to demand plans. However, if you do not know the demands, you cannot adjust the equation (El-Sakty and Osama 2015). Then we need real-time visibility into what is going on throughout the network, so you can instantly pivot to whichever of the scenarios is taking place. It is a fundamental shift in the way supply.

1.3.1 Black Swan Events

A "*black swan*" is an event which cannot be foreseen, but when it occurs it changes the whole understanding of the market and has major effect (Taleb 2020). Further, the COVID-19 was perceived as a white swan event due to the certainty of the issues of the virus outbreak. However, when it became a global pandemic it turned out to be black swan event as the nations were not aware of the impact of the COVID-19 on the supply chain and how industries can operate in the era of the COVID-19 outbreak (Siegenfeld et al. 2020). An ancient saying that assumes black swan did not prevail—a saying that became re-explained to teach a different lesson after the first European experience. The events with a very low probability of occurring and consequently are difficult to predict, however, have a potentially disastrous impact. These events are often referred to as black swans (Taleb 2007). A black swan is an event that has an almost zero likelihood of occurring, but, if it does, can have a major impact (Liu and Wang 2011). Black swan events are difficult to define because they may have never occurred. COVID-19 can be considered as the black swan event that compels many industries and companies to re-assess and change their GSCM. COVID-19 surprises the world as normal black swan events (Kilpatrick and Barter 2020).

1.4 Global Supply Chain: Disruption

The COVID-19 outbreak is indeed disrupting global supply chains and many companies have a Tier 1 or Tier 2 vendor that has been impacted directly or indirectly by the virus (Deloitte 2020). Impact of the COVID-19 pandemic is felt across the supply chain, leading to labor shortages and travel restrictions which are unavoidable (Queiroz et al. 2020). Moreover, a black swan event occurs in an organization, and its impact is felt all over the supply chain (Scheibe and Blackhurst 2017). Moreover, having the right information about the business operations can predict the unexpected events. Furthermore, in global crisis, the supply chain visibility is crucial and becomes more critical for flawless business operations (Sreedharan et al. 2020). The global supply chain disruption highlighted tremendously the ripple effect and resilience importance in global operations. The paper "Exploring Supply Chain Network Resilience in the Presence of the Ripple Effect" by Li and Zobel (2020) highlights how the ripple effects can create disorder across the supply chain network. Moreover, to understand the supply challenges in various industry in the VUCA world, the work reviewed the various scholarly articles and tabulated the findings in Table 1.1.

Based on the review of Table 1.1, GSCM consists of different practices, which are discussed in the following sections.

Table 1.1 Review of supply chain activities in the different industries

S. No.	Authors	Approach	Supply chain Practices	Application
1	Badakhshan et al. (2020)	Network optimization	Supply chain collaboration	Financial management
2	Brevik et al. (2020)	Simulation	Supply chain network	Broiler production
3	Flores and Villalobos (2020)	Agro supply chain	Supply chain design	Food production
4	Ren et al. (2020)	Circular economy	Supply chain network	Plastic cycling
5	Azzamouri et al. (2019)	Network optimization	Supply chain Co-ordination	Fertilizer production
6	Bottani et al. (2019)	Multiple-sourcing policy	Supply chain design	Food supply
7	Kellner and Utz (2019)	Posteriori approach	Supply chain collaboration	Portfolio management
8	Lechner and Reimann (2019)	Circular economy	Supply chain design	Product recycling
9	Priore et al. (2019)	Material procurement	Supply chain collaboration	Inventory management
10	Schätter et al. (2019)	Resource allocation	Supply chain modelling	Robust system
11	Sharahi et al.	Fuzzy optimization	Supply chain design	Natural gas supply chain
13	Zhao et al. (2019)	Optimization techniques	Supply chain transaction	Resilience
14	Sreedharan et al (2019)	Pharma supply chain	Supply chain design	Pharmaceutical company

1.4.1 Cargo Carrier Assessments

Wide ranging of evaluation is necessary for delivering cargo on time at affordable cost. It should thoroughly analyze before getting into the transportation partner. We should create a framework as valuable tools while performing cargo carrier examination (Araz et al. 2020).

1.4.2 Geopolitical Instability

Supply chain disruptions are generally caused by trade wars, protests, etc. If a particular city or country encountered unrest, it will produce various issues in supply chain. Civil unrest leads to loss of communications, and time delays in supply chain problems. The size and time duration of these events are unpredictable (Helbig et al. 2016).

1.4.3 Cultural Research

Supply chain decisions are made with background information where political instability is common. Research has to be conducted on climate of the geographic area where we get the resource from. When a chance of instability is found then that region should be heavily considered. If there is no option, we have to develop a strong contingency to avoid worst-case scenario (Alghababsheh and Gallear 2020).

1.4.4 Price Hikes

Cost of raw materials influences the component value in the supply chain. Also, market price is generally unpredictable which creates threat to the supply chain (Tang 2006).

1.4.5 Market Exploitation

Global market does not affect an individual person but impacts everyone. The best way to reduce threat related to rising prices is to understand the market's trend. Having knowledge of market trends allows us to decide when we have to lower inventory raw materials and purchase more of that material. Research and analysis suggest as a report regarding the market trends and we have to act accordingly for the future works (Liu et al. 2016).

1.4.6 Cyber-Attacks

Cyber-attacks and computer hacking impacts supply chain events over the last five years. Immeasurable amount of data flows through a supply chain every day. Therefore, industries must be concerned over this area to avoid supply rain risk. New technologies are invented to safeguard against cyber threats. To protect against natural disaster risk, we have to backup suppliers from various geographical region (Mensah and Merkuryev 2014). Also, we have to make sure about the contingency plans for distribution centers and transportation routes in case of natural disaster. Distribution network should be set up in a place where items can move from various global regions. Making a more dynamic in adapting new technologies without lagging in security and safety is the most required paradigm in the actual market scenario (Wolden et al. 2015; Rejikumar et al. 2019). Global extravagance lies in delivering qualitative products at right time ensuring the safety and security from any kind of miscellaneous cyber-attacks as the current situation affected the supply chain system in terms of delivery, marketing, valuations, security, distribution, and customer engagement.

1.4.7 Transportation Failure

Supply chain disruptions in transportation are regulation changes, shipping damage, cargo theft, and transit delay. Now, new logistics system and transportation services are creating more competitiveness to today's challenge (Liu et al. 2016). COVID-19 situation forced to put travel restrictions and in some worst cases travel ban has happened for all types of goods and commodities from highly pandemic-affected countries, leveraging the supply chain in air (Ivanov and Dolgui 2020). With these restrictions, few of the consignments missed their delivery timeline and forced many trading companies to withdraw many international product transportations leading to major setback in supply chain market.

1.5 Impact of COVID-19 on the Supply Chain Practices

To face the black swan events the supply chain has to be both positive in agility angle and able to absorb the sudden spike; and avoid disruption and able to recover in a short span of time (Ivanov 2020). The business which had shown few signs of reviving has again been delayed and the financial stress in the industry continues. Things like material shortages and increased rates on account of the shortage are impacting the decision-making of foundries. Since component inventories for assemblies that are sourced from China are affected, the schedules which were to resume may not happen and projects which were waiting for commissioning with support from China are on a standstill due to travel restrictions, and funds crunch has initiated (Kilpatrick and Barter 2020). The short-term impact of supply chain mainly lies in the lack of personal flexibility, difficulty in recovering production, and transport restrictions. Also, the degree of disruptions increases as the time to recovery increases in the supply chain network. Because the majority of the manufacturing hub are centered in China, the pandemic can cripple the supply of the necessary components to other nations (Kilpatrick and Barter 2020). Customer-responsive strategy is commonly used among businesses in global supply chain. Focusing mainly on customer, that is considering customer perceptive, helps to understand the difficulty in customer values, and solutions to these challenges are implemented and businesses have increased fame on cost-reduction efforts (Raghuram et al. 2020).

- To move forward, updating the latest trends, monitoring the data insight, and grasping the new customer behavior will be the key to successful marketing in supply chain.
- Advertisers can reach the customer in digital platforms like TV and online as users are staying at home. They can invest more in advertising through television and social networking due to decrease in in-store traffic.

Logistics firms are directly affected by the COVID-19 pandemic which is involved in storage, movement, and flow of goods. It helps to deliver the products to customers across borders. Supply chain disruption occurs when these are affected by this current pandemic. Logistics generally facilitate trade and commerce to help business (Albors-Garrigos 2020).

1.5.1 Exit Strategy for the GSCM

Exit strategy will be considered when developing a commodity strategy as a front-end activity. Further, it should be included in the acquisition documents and contractual terms and conditions where possible. We need a strategy which is consistent with overall sourcing strategy. Also, exit strategy will include the needs in the contract and it is placed at the start of a supplier relationship. Exit strategies should be reviewed annually and there are several considerations to be made when developing an exit plan, including:

1. Pursue assistance requirements
2. Data safety
3. Knowledge and documentation transfer
4. Expense
5. Manpower.

According to their own merits, contract relationship should be considered. The COVID-19 outbreak may have deepened the pushback against globalization among the different countries that were building a new supply chain partnership with multi countries. Further, the COVID outbreak questions the stability of the global supply practices and promotes many countries to move toward import substitution ("reshoring") (Polatidis et al. 2018). In this pandemic time, considering the exit strategy in global supply chain is very important in terms of supplier relationship.

1.5.2 Rethinking the Global Value Chain

In global value chain, the COVID-19 outbreak act as a wake-up call for companies and improve their actions and ready for flexibility in future unpredictable events.

Chinese manufacturers play major role in the supply chain of other companies. The impact of COVID-19 in Chinese manufacturers is felt around the world because it plays a central role in supply chain. Industries like electronics, automobiles, pharmaceuticals, and other consumer products like surgical gowns and masks had an acute disruption. Supply chain designs are interdependent with every sector and the management should think and act to get through this complex structure (Dolgui and Ivanov 2020). In recent years, supply chains have become progressively global. There is an increase in the number of tradable products and services. The products produced remotely from the market, which are consumed, are determined as tradability. Transportation cost and perish ability are the two main factors in tradability. Many companies shifted to global sourcing by taking advantage of cost reduction in labor, materials, and lands in low-cost region. Goods with high value, depending on the size and shipment cost, were produced in low-cost region. Subcontracting is an important factor in the growth of the supply chain. It is adopted for numerous reasons, which include specialists required in manufacturing process; increase in knowledge of components; and flexible capacity depending on the demand. Subcontracting results in rooted tiring of supply chains that form a multistage production network. Suppliers form their own network of suppliers having many tiers. It is extremely difficult for companies to have clear visibility over the suppliers. This is where the major disruption occurs. Earlier we had many black swan events, but the disruption caused by COVID-19 is different from the other. Country risk rate are unpredictable in this event. No one could imagine that the world's second largest economy shut down external logistics and went offline. Because of supply chain tiring there are delays in shipping, and organizations are highly dependent and vulnerable to this uncertainty.

1.5.3 Re-Assessing Supply Chain Risk

Supply chain risks associated with segmentation and globalization are ignored over the past few years. But the current situation shows the risks and their impacts. Joining global multilevel supply networks and lean production leads to crises. Managers should understand their supply chain's vulnerabilities based on the above fact and possible actions have to be taken. They are:

1.5.3.1 Consider Regionalization

The U.S.–China trade war has already made regionalization of manufacturing back on track. The production houses are considering localizing the supplier to avoid disruption. Toyota from Japan and other Chinese manufacturers pioneered in lean production, and collocated the suppliers nearby. To apply lean and on-time production methods for efficient logistics and transport, many companies are following this.

1.5.3.2 Develop Second Sources or Additional Safety Stocks

Developing second source or saving additional safety stocks will increase the cost. We need to think as a long-term benefit for supply chain resilience. There are difficulties in concentrating the manufacturer or supplier and resources location. It is important to calculate how much safety stock is needed depending on the circumstances. In complex manufacturing process, we have to develop proper alternative which needs certification. We face serious challenges when the suppliers are delivering sources with unique capabilities. For example, Semiconductor Manufacturing Company in Asia has the largest global market for most of the advanced chips, and majority of the smartphone manufacturers rely on this company for advanced chips (Carson and Hung 2014). Asian countries are the major contributor for the world's semiconductor integrated circuit. In turn, some Asian countries are dependent on Dutch supplier for lithography systems and the corresponding supplier is dependent on a factory in Europe for its optical engine. Because of the critical dependency, some Asian government started their own initiative such as making in their own countries. Developing an alternate source of supply will reduce risk in supply chain. Long-term purchasing commitment will make new supplier to match the price with the existing customer. In developing the safety stock and resources, buying from higher cost producer will not satisfy the manufacturers.

1.5.3.3 Re-Think Scale and Product Mix

Many companies focused on factories that manufacture variants in single platform. Handling a wider mix of products within same factories will give regionalized production when disruption occurs. This approach is more popular in industries like consumer goods and electrical appliances. Further, IMF forecasts that U.S. GDP for 2020 can be 8% lower than previous year and recovery in 2021 is yet to be fully explored (Kranke 2020).

1.5.3.4 Challenges Facing Supply Chain Practices

Every organizations need to follow safety measures, such as wearing masks, social distancing, and a strong response to avoid surges which threaten supply chain. The infection results in localized knockdowns and business interruptions. Supply chain organizations understood that Tier 1 suppliers are not enough. Much of this was due to faulty assumptions regarding Tier 1 suppliers and possibly a degree of wishful thinking. Also, the supply was affected by shutdown of Tier 2 and Tier 3 suppliers in the virus spreads region and in other parts of the world. Therefore, it is pivotal for organizations to evaluate their supply chain activities in detail, to detect the faults (Handfield and Nichols 2004). Such practices will facilitate effective action by the organization when the downstream supply chain is threatened. Also, companies should stop focusing not only on supply chain visibility but also in resource investment and study supply chain difficulties. It is a tedious process particularly in large organizations. The outbreak like COVID-19 taught us that if companies did not pay more attention in supply chain crisis, it will cause weaknesses in supply chain. Precautions method is not only related to COVID-19, but organizations need ample data to predict the impact of other unpredicted events such as:

- Commercial wars
- Calamities
- Civil disorder
- Protests.

1.6 Summary of the Findings

As the COVID-19 outbreak has disrupted the supply chain practices, it is instrumental to develop effective approach to deal with the COVID-19 crisis for supply chain resilience in the VUCA world. Table 1 has highlighted various functions in the supply chain and approaches used to manage the supply chain practices.

- The COVID-19 outbreak not only affects the supply chain but also affects societies and economies at their core. Businesses are eager to boost productivity and demand in the global supply chain.
- The characters of VUCA are the challenges that we face due to COVID-19 outbreak, but the successful leadership will change the model to vision, understanding, clarity and adaptability.
- The black swan event is compared with the current pandemic. The pandemic is unexpected and hard to predict because it is beyond the normal expectations.
- The pandemic control is a collective effort and support to lower capacity countries.
- The supply chain risk determines business nature, difficulty of the chain, and number of products included.
- The three components of global supply chain are: marketing, logistics, and supply.
- When a particular firm is disrupted, its effects are often felt throughout the supply chain network because they are interlinked.
- In economic growth, consumer sector contributes the most. In this pandemic, consumer sector was under pressure and the impact is likely to be short term.
- The effects of COVID-19 pandemic are difficult to access and model, but businesses begin to reduce the supply chain distributions.
- Insights about the pre and post COVID-19 impacts on supply chain market and the major setbacks for trading and marketing industries to meet up their customer engagements and balancing their workforce and deliverables are researched.
- Activating raw materials substitutions and developing alternative supply chain scenarios result in change in pricing strategies.
- Frequent financial modeling and scenario planning is required to protect growth and increase profitability without any resilience.

- Take long-term considerations in business models as a result of pandemic.
- Furthermore, GSCM consists of different supply chain practices, such as supply chain network; supply chain co-ordination; supply chain design; supply chain collaboration; supply chain design; supply chain modeling; and supply chain transaction, that require continuous monitoring for robust operations without deviation.

Further, many firms are mostly dependent on China because of availability of domain experts and abundant resources for production. Therefore, developing a contingency plan to avoid supply chain disruption and also to design future supply chain model, decision makers should include new measures, including responsiveness, resilience, and re-configuring apart from quality, delivery, and cost.

1.7 Conclusion

As the world is affected by the COVID-19 outbreak, the major challenges for organization and industries is to co-ordinate the supply chains practices globally. At this juncture, the supply chain visibility is instrumental as it offers more clarity to take effective decision making in the crisis. Furthermore, the supply chain visibility is instrumental in developing a feasible exit strategy for the organization and minimize the uncertainty in the business operations. Moreover, the data collection is critical for supply chain practices to make a map for the supply chain transactions to reduce the ambiguity and make the practices more resilient for future disruptions. Furthermore, the COVID-19 impact on the GSCM is a black swan event, making it difficult to predict and difficult to contain as it is highly uncertain. Such black swan events drive the company to adopt exit strategy which are instrumental for the supply chain recovery and it is vital for the organization. Also, one has to ensure that the global value chain has to be assessed without any issues and to re-assess the supply chain risk for the global operations. Adding to this, various challenges arise in the supply chain practices which need to minimize global supply chain operations, and it is clearly explained in the following chapters.

References

Ali, S. S., Kaur, R., Persis, J, Saha, R., Murugan, P., & Sreedharan, R. (2020). Developing a hybrid evaluation approach for the low carbon performance on sustainable manufacturing environment. *Annal Operation Research*, (Ahead of print) https://doi.org/10.1007/s10479-020-03877-1.

Azzamouri, A., Bara, N., Elfirdoussi, S., Essaadi, I., Fontane, F., & Giard, V. (2019). DSS approach for heterogeneous parallel machines scheduling considering proximate supply chain constraints. *International Journal of Production Research*, (Ahead of Print).

Albors-Garrigos, J. (2020). Barriers and enablers for innovation in the retail sector: co-innovating with the customer. A case study in grocery retailing. *Journal of Retailing and Consumer Services, 55*, 102077.

Alghababsheh, M., & Gallear, D. (2020). Socially sustainable supply chain management and suppliers' social performance: The role of social capital. *Journal of Business Ethics, 163*, 125–150. https://doi.org/10.1007/s10551-018-3993-0.

Anderson, D. L., Britt, F. F., & Favre, D. J. (2007). The 7 principles of supply chain management. *Supply Chain Management Review, 11*(3), 41–46.

Araz, O. M., Choi, T. M., Olson, D. L., & Salman, F. S. (2020). Role of analytics for operational risk management in the era of big data. *Decision*. https://doi.org/10.1111/deci.12451.

Badakhshan, E., Humphreys, P., Maguire, L., & McIvor, R. (2020). Using simulation-based system dynamics and genetic algorithms to reduce the cash flow bullwhip in the supply chain. *International Journal of Production Research*, (Ahead of Print).

Bennett, N., & Lemoine, G. J. (2014). What a difference a word makes: Understanding threats to performance in a VUCA world. *Business Horizons, 57*(3), 311–317.

Bottani, E., Murino, T., Schiavo, M., & Akkerman, R. (2019). Resilient food supply chain design: Modelling framework and metaheuristic solution approach. *Computers & Industrial Engineering, 135*, 177–198.

Brevik, E., Lauen, A. Ø., Rolke, M. C. B., Fagerholt, K., & Hansen, J. R. (2020). Optimisation of the broiler production supply chain. *International Journal of Production Research*, (Ahead of print).

Carson, M. T., & Hung, H. C. (2014). Semiconductor theory in migration: Population receivers, homelands and gateways in Taiwan and Island Southeast Asia. *World Archaeology, 46*(4), 502–515.

Choi, T. Y., Rogers, D., & Vakil, B. (2020). Coronavirus is a wake-up call for supply chain management. https://hbr.org/2020/03/coronavirus-is-a-wake-up-call-for-supply-chain-management. Retrieved 20 August 2020.

Deloitte. (2020). COVID-19: Managing supply chain risk and disruption. https://www2.deloitte.com/global/en/pages/risk/articles/COVID-19-managing-supply-chain-risk-anddisruption.html. Retrieved 20 August 2020.

Dolgui, A., & Ivanov, D. (2020). Exploring supply chain structural dynamics: New disruptive technologies and disruption risks. *International Journal of Production Economics*.

El-sakty, k. G., & Osama, E. (2015). Improving supply chain performance using 'vuca' drivers. *Journal of Supply Chain Management: Research and Practice, 9*(1), 1–10.

Flint, D. (2004). Strategic marketing in global supply chains: Four challenges. *Industrial Marketing Management, 33*(1), 45–50. https://doi.org/10.1016/j.indmarman.2003.08.009.

Flores, H., & Villalobos, J. R. (2020). A stochastic planning framework for the discovery of complementary, agricultural systems. *European Journal of Operational Research, 280*(2), 707–729.

Guan, D., Wang, D., Hallegatte, S., Davis, S. J., Huo, J., Li, S., & Cheng, D. (2020). Global supply-chain effects of COVID-19 control measures. *Nature Human Behaviour*, 1–11.

Handfield, R. B., & Nichols, E. (2004). Key issues in global supply base management. *Industrial Marketing Management, 33*(1), 29–35. https://doi.org/10.1016/j.indmarman.2003.08.007.

Hasani, A., Zegordi, S. H., & Nikbakhsh, E. (2012). Robust closed-loop supply chain network design for perishable goods in agile manufacturing under uncertainty. *International Journal of Production Research, 50*(16), 4649–4669.

Helbig, C., Gemechu, E. D., Pillain, B., Young, S. B., Thorenz, A., Tuma, A., et al. (2016). Extending the geopolitical supply risk indicator: Application of life cycle sustainability assessment to the petrochemical supply chain of polyacrylonitrile-based carbon fibers. *Journal of Cleaner Production, 137*, 1170–1178.

Ivanov, D., & Dolgui, A. (2020). Viability of intertwined supply networks: Extending the supply chain resilience angles towards survivability. A position paper motivated by COVID-19 outbreak. *International Journal of Production Research, 58*(10), 2904–2915.

Ivanov, D. (2020). Viable supply chain model: Integrating agility, resilience and sustainability perspectives—lessons from and thinking beyond the COVID-19 pandemic. *Annals of Operations Research.*

Jia, J., & Hu, Q. (2011). Dynamic ordering and pricing for a perishable goods supply chain. *Computers & Industrial Engineering, 60*(2), 302–309.

Kranke, M. (2020). IMF-world bank cooperation before and after the global financial crisis. *Global Policy, 11*(1), 15–25.

Kellner, F., & Utz, S. (2019). Sustainability in supplier selection and order allocation: Combining integer variables with Markowitz portfolio theory. *Journal of Cleaner Production, 214,* 462–474.

Kilpatrick, J., & Barter, L. (2020). *COVID-19: Managing supply chain risk and disruption.* Deloitte. https://www2.deloitte.com/content/dam/Deloitte/ca/Documents/finance/Supply-Chain_POV_EN_FINAL-AODA.pdf. Accessed November 2020.

Koberg, E., & Longoni, A. (2019). A systematic review of sustainable supply chain management in global supply chains. *Journal of Cleaner Production, 207,* 1084–1098.

Lechner, G., & Reimann, M. (2019). Integrated decision-making in reverse logistics: An optimisation of interacting acquisition, grading and disposition processes. *International Journal of Production Research,* (Advance online publication).

Li, Y., & Zobel, C. W. (2020). Exploring supply chain network resilience in the presence of the ripple effect. *International Journal of Production Economics, 228,*.

Liu, Y., & Shasha, Wang. (2011). Research on collaborative management in supply chain crisis. *Procedia Environmental Sciences, 10,* 141–146.

Liu, Z., Prajogo, D., & Oke, A. (2016). Supply chain technologies: Linking adoption, utilization, and performance. *Journal of Supply Chain Management, 52*(4), 22–41.

Manuj, I., Esper, T. L., & Stank, T. P. (2014). Supply chain risk management approaches under different conditions of risk. *Journal of Business Logistics, 35*(3), 241–258.

Mensah, P., & Merkuryev, Y. (2014). Developing a resilient supply chain. *Procedia-Social and Behavioral Sciences, 110,* 309–319.

Mentzer, J. T., Myers, M. B., & Cheung, M.-S. (2004). Global market segmentation for logistics services. *Industrial Marketing Management, 33*(1), 15–20. https://doi.org/10.1016/j.indmarman.2003.08.005.

Musa, A., Gunasekaran, A., & Yusuf, Y. (2014). Supply chain product visibility: Methods, systems and impacts. *Expert Systems with Applications, 41*(1), 176–194.

Packowski, J. (2014). *Lean supply chain planning: The new supply chain management paradigm for process industries to Master today's VUCA world.*

Peidro, D., Mula, J., Poler, R., & Lario, F. C. (2009). Quantitative models for supply chain planning under uncertainty: A review. *The International Journal of Advanced Manufacturing Technology, 43*(3–4), 400–420.

Polatidis, N., Pavlidis, M., & Mouratidis, H. (2018). Cyber-attack path discovery in a dynamic supply chain maritime risk management system. *Computer Standards & Interfaces, 56,* 74–82.

Priore, P., Ponte, B., Rosillo, R., & de la Fuente, D. (2019). Applying machine learning to the dynamic selection of replenishment policies in fast-changing supply chain environments. *International Journal of Production Research, 57*(11), 3663–3677.

Queiroz, M. M., Ivanov, D., Dolgui, A., & Wamba, S. F. (2020). Impacts of epidemic outbreaks on supply chains: mapping a research agenda amid the COVID-19 pandemic through a structured literature review. *Annals of Operations Research,* 1–38.

Raghuram, P., Sandeep, P., Sreedharan, V. R., & Saikouk, T. (2020). Development of a supply chain risk mitigation index for distillery. *The TQM Journal,* (Vol. ahead-of-print No. ahead-of-print). https://doi.org/10.1108/TQM-01-2020-0008.

Rejikumar, G., Sreedharan V., R., Arunprasad, P., Persis, J., & Sreeraj, K. M. (2019). Industry 4.0: Key findings and analysis from the literature arena. *Benchmarking: An International Journal, 26*(8), 2514–2542.

Ren, H., Zhou, W., Guo, Y., Huang, L., Liu, Y., Yu, Y., et al. (2020). A GIS-based green supply chain model for assessing the effects of carbon price uncertainty on plastic recycling. *International Journal of Production Research, 58*(6), 1705–1723.

Scheibe, Kevin P., & Blackhurst, Jennifer. (2017). Supply chain disruption propagation: A systemic risk and normal accident theory perspective. *International Journal of Production Research, 56*(1–2), 43–59.

Schätter, F., Hansen, O., Wiens, M., & Schultmann, F. (2019). A decision support methodology for a disaster-caused business continuity management. *Decision Support Systems, 118,* 10–20.

Siegenfeld, A. F., Taleb, N. N., & Bar-Yam, Y. (2020). Opinion: What models can and cannot tell us about COVID-19. *Proceedings of the National Academy of Sciences, 117*(28), 16092–16095.

Sreedharan, V. R., Kek, V., Dhanya, M., Anjali, S., & Arunprasad, P. (2020). Understanding the role of logistics in humanitarian operations: key findings and analysis from literatures. *International Journal of Logistics Systems and Management, 36*(4), 463–494.

Sreedharan, V. R. S. R., Kannan, S. S. P. A., & Trehan, R. (2018). Defect reduction in an electrical parts manufacturer: a case study. *The TQM Journal, 30*(6), 650–678.

Sreedharan, V. R., Kamala, V., & Arunprasad, P. (2019). Supply chain risk assessment in pharmaceutical industries: an empirical approach. *International Journal of Business Innovation and Research, 18*(4), 541–571.

Srinivas, S. S., & Sreedharan, V. R. (2018). Failure analysis of automobile spares in a manufacturing supply chain distribution centre using Six Sigma DMAIC framework. *International Journal of Services and Operations Management, 29*(3), 359–372.

Sumner, A., Hoy, C., & Ortiz-Juarez, E. (2020). *Estimates of the impact of COVID-19 on global poverty.* WIDER Working Paper 2020/43.

Taleb, N. N. (2020). Skin in the game: Hidden asymmetries in daily life. *Random House Trade Paperbacks.*

Taleb, N. (2007). *The black swan: The impact of the highly improbable.* New York: Random House Inc.

Tang, C. S. (2006). Perspectives in supply chain risk management. *International Journal of Production Economics, 103*(2), 451–488.

Wolden, M., Valverde, R., & Talla, M. (2015). The effectiveness of COBIT 5 information security framework for reducing cyber-attacks on supply chain management system. *IFAC-PapersOnLine, 48*(3), 1846–1852.

Zhao, K., Scheibe, K., Blackhurst, J., & Kumar, A. (2019). Supply chain network robustness against disruptions: Topological analysis, measurement, and optimization. *IEEE Transactions on Engineering Management, 66*(1), 127–139.

Chapter 2
Challenges Faced and Preparedness of FMCG Retail Supply Chain During COVID-19

Ravi Shankar Bhakat and Md. Zahir Uddin Arif

Abstract Coronavirus pandemic or COVID-19 outbreak has created huge disruptions in the supply chains of goods and commodities. Continuous reconfiguration of supply chain is the new theme in the post-COVID world for all the FMCG retailers. A series of lockdown have caused much more serious implications for the businesses and firms. Today, FMCG producers and retailers need more flexibility than ever in their global and local supply chain process to be successful. A resilient supply chain for FMCG goods is the need of the hour which requires understanding of any contingencies. In addition, after the crisis traditional modern retailers inside shopping malls are affected, and there is a pressing need to re-emphasize on digital platforms and partners. Due to behavioral changes, proper planning and assessment of consumer demand, logistics infrastructure, warehousing capabilities etc. are quintessential for managing the stockpile of critical FMCG items.

Keywords FMCG distribution · Digital supply chains · Retailers · Critical goods

2.1 Introduction

Coronavirus pandemic or COVID-19 outbreak has created huge disruptions in the supply chains of goods and commodities. Deloitte website report states that COVID-19 shocked the world and twisted the normal life across the globe. The disruptive phenomenon started with coronavirus outburst in Wuhan, capital city of Hubei province in central part of China. Post December 31, 2019 with the first case of the potential epidemic, it turned into a global pandemic by the end of February 2020 with almost 100,000 people affected and 3,000 mortality cases (Ivanov 2020).

R. S. Bhakat (✉)
Management Development Institute Murshidabad (MDIM), Murshidabad West Bengal,, India

Md. Z. U. Arif
Faculty of Business Studies, Department of Marketing, Jagannath University, Dhaka 1100, Bangladesh

A. R. Sakthivel et al. (eds.), *Managing Supply Chain Risk and Disruptions: Post COVID-19*, Management and Industrial Engineering,
https://doi.org/10.1007/978-3-030-72575-4_2

Coronavirus disease 2019 (COVID-19) has created a crisis across the globe. Businesses realized in a hard way that supply chains activities are the core processes of global economy. Societies and communities across the world faced unforeseen problems due to lack of essentials supply during the COVID outbreak. A series of lockdown have caused much more serious implications for the businesses and firms. Without a doubt, COVID-19 has re-emphasized the need of supply chain for our economy and society. To control the effects of such a crisis on the individual supply chain, an effective management must quickly implement appropriate measures (Yu et al. 2020).

Compared to all businesses, retail has been the most affected business by the pandemic. Unlike other sectors, retail supply chains are daily affairs coordinating various routes and schedules. Till now, this has been made possible with the option of providing opportunities for replenishment for large and small retail stores. Further adaptation in a strenuous environment and new market conditions would test the survival of retail chains. Global supply chains have been disrupted due to loss of labor and raw material inputs, creating ripple effects that cross national boundaries (Yu and Aviso 2020).

However, even before the pandemic, supply chain in retail has always been unique. Fast-moving consumer goods (FMCG) is the article of daily use having a high consumption frequency. FMCG supply chain managerial decisions focus on finding out when to order and how much to order. Keeping up the optimum flow of goods is usually a priority of FMCG supply chain. The complexity, expected efficiency and cost is peculiar for retail supply chain and logistics (Huber et al. 2017).

Numerous SKUs require multiple mode of transportation and warehousing of goods; at the same time ensuring adequate stock. All these are efficiently tackled with a high level of software support systems across the various coordinated activities. Moreover, the nature of the retail network is also influenced greatly by the number, size and objective of the retail partners along with the mode of transports (Mihajlović and Trajković 2020).

2.2 Behavioral Change and Post Crisis Supply Chains

When it comes to retail supply chain, the implications are more severe. This demands an additional push for the supply chain management after the pandemic situation. At the same time, special provisions should be made to keep up the logistics and allied activities for future crisis (Hobbs 2020).

When the essential goods consumption is growing continuously due to behavioral change in the market, FMCG products need to be given that extra focus in terms of production, procurement and processing of orders. FMCG retailers need to assess in real time pertinent to the availability of critical low- or high-value goods and products. There would be high impact on the revenue streams of retail businesses if various bottlenecks related to order quantity, backend capacity and price volatility are not efficiently managed.

FMCG goods as the major chunk of essential goods needed extra emphasis and consideration. The movement of FMCG goods has undergone a series of transformation due to the challenges posed by changing consumers' needs. The situation of COVID-19 is turning out to be a natural disruptive phenomenon that is forcing many FMCG firms, viz., Unilever, P&G, ITC and similar companies, and even the whole industry to redesign and realign their supply chain structure and processes.

When we look at the overall potential, the global FMCG market is expected to reach 15 trillion dollar in next five years, growing consistently despite major challenges. E-commerce growth coupled with traditional brick and mortar outlets is catalyzing the growth of this sector. Even during the lockdown period, offline outlets of major retailers have started acting as the backend support system of the online commerce activities. Major shopping mall was temporarily transformed into supply chain networks and hubs.

2.3 Challenges for Retailers and Partners

COVID-19 has created a separate impact on suppliers and buyers in terms of behavioral change. Stakeholders of the supply chain must take all those compulsory habitual shifts into account. Prima facie, we can say that customers have started giving unforeseen priority to health and safety over price and easy accessibility. Retailers of FMCG products must necessarily readjust strategic and operational activities to ensure the health and safety. One evident way is to minimize the contacts and reduction of various human touch points. Going by the majority of health and safety conscious suppliers and buyers, it must be ensured that retailers and FMCG firms take special steps to enhance the sanitation and hygiene of supply chains. In the pursuit of velocity and efficiency in FMCG retail, the probability of fluctuation grows (Scheibe and Blackhurst 2017).

2.3.1 Panic Buying

COVID created a sense of panic among the consumers due to unprecedented change in the society status quo. The end-consumers felt the impact of emergency stock up of essentials during self-isolation. Therefore, observing the essential goods and commodities can provide first-hand knowledge of dealing with respect to the shelf life. For fastest moving goods with rather short shelf life, retailers have to tackle up the challenges of out-of-stock crisis due to unprecedented panic buying from consumers. Supply chain partners, therefore, should rework on the traditional processes in terms of planning, forecasting, replenishment of essential and low shelf life products. Proper planning and assessment of consumer demand, logistics infrastructure, warehousing capabilities etc. are quintessential for managing the stockpile of critical FMCG items.

Retailers and partners should take certain steps to reduce the impact of panic buying and subsequent stockpiling activities by consumers. All the chances of future crisis need careful consideration. The understanding of the motivations behind such behaviors can also mitigate the negative effect of such behavior at retail store level.

However, putting a limit to purchases of critical items may deal with panic buying temporarily. But, there is a need to explore the fundamental impact on the retail setups in case of chronic pandemic crisis and habitual panic buying (Hobbs 2020).

2.3.2 Procurement

Procurement and sourcing of commodities, goods and branded products poses a serious challenge to retailers. Due to lockdown, there is a need to match the availability and capacity with respect to critical products. There must be revisit of traditional contingency plans and backups which demand reviewing of supplier's relationships and creating alternatives for supply chain emergency-like situations. This would ensure that the possible risks are mitigated optimally. In addition, individual supplier's risk assessment can be done in terms of their quick ability to meet the requirements of retailers. At the same time, strengthen the procurement aspect of supply chain while focusing upon the right quality at right time at right place.

The rapid replenishment of inventories is not sufficient if there is a violation with the appropriate quality standards of goods, commodities or products. To ensure the quality, packaging and transportation need equal emphasis for FMCG products. All the partners of supply chains therefore need to be made aware about the inability to tradeoff availability with quality. Regular quality check and monitoring of flow of goods are critically essential. Information technology for real-time awareness about the supplier's activities and schedule need to be developed to meet the required objectives.

2.3.3 Disintermediation and Digital Shift

After the crisis, the traditional modern retailers inside shopping malls are affected, and there is a pressing need to re-emphasize on digital platforms and partners. The coronavirus lockdown has convinced the FMCG suppliers and distributors to transcend their comfort zones in terms of old supply chains. Due to lack of human resource in shifts, many activities and schedules are put to halt. Disintermediation has put digital supply chain on priority to ensure continuous supply of essential goods, viz., staples to households. Digitization was earlier a revolutionary and proactive measure for firms, but now it has become a reactionary and must have tool. Most FMCG companies have managed quick tie-ups with delivery companies.

The ongoing crisis of COVID also emphasizes on revisiting the role of intermediaries. The roles and responsibilities must be reassessed to tackle with the emergency-like situation. While making it more end-consumer-centric, retailers must try to shorten the supply chain by engaging actively in the distribution systems, be it online retailing model or peer-to-peer systems.

The re-emergence of online model with increased number of users shopping online for essential commodities including FMCG and groceries creates a new shift in the consumer behavior. It is therefore essential to focus on the supporting logistics, order fulfillment, third-party logistics, procurement management and customer service.

COVID-19 is becoming a catalyst in the process of firms restructuring their global and local supply chain strategy. It also increased the adoption of digital supply chain models across the consumer goods and industrial goods. A series of small and concrete actions taken in a coherent manner can be done to tackle the immediate challenge. But to make a provision for long-term solution, there is a need for much overdue transformational shift in the traditional supply chains.

2.4 Preparing for a Resilient FMCG Supply Chain

The design of a resilient supply chain for FMCG goods requires the understanding and development of adaptive processes with respect to external changes. When the unregulated distribution system is affected, established big retailers and suppliers must focus upon customer orientation to keep supply chains intact. More and more retailers prefer to get their supply chain localized in their respective region. Flexibility can be considered as the goal of making a supply chain able to alter conditions to meet an unanticipated situation (Chuu 2011).

2.4.1 Empowering the FMCG Retail Stakeholders

Educating the employees and partners on precautionary methods to tackle COVID-19 could be a good start. For FMCG retailers, empathy and caring of stakeholder must be the first and foremost objective. Firms and their partners must increase awareness among the in-house staffs and external workforce including key suppliers and partners. The continuous monitoring of the people in front end and customer-facing profiles should regularly update the health records in real time. Based upon the age, immunity and potential exposure-related aspects, firms can groom them for an enriched job profile. Suitable communication strategies can be deployed by FMCG retail supply chain stakeholders to reduce the impact of end-consumers panic-buying activities (Hobbs 2020).

Regionalization or localization is another trend for empowering the FMCG retail stakeholders. There could be high priority decisions related to enhanced distribution of products procured locally in the vicinity of the consumption hubs. Logistics decisions of FMCG supplies including transport, warehousing, handling and packaging should be mutually reassessed with all stakeholders. This must be done to empower the partners and make the supply chain equipped with efficient and agile solutions (Choi 2020).

At the same time, retailers can also look for exploring the parallel supply chains, thereby reducing the time to replenish the forecasted or actual demands. Close interaction between a 3PL and its logistics collaborators also facilitates information sharing, preparation and planning, thereby helping to build resilience (Liu and Lee 2018).

2.4.2 Demand Planning

When some sectors have seen negligible or no demand post COVID disruption, the FMCG sector is one exclusive sector which has faced delayed demand, decreased demand and exponentially high demand. Single source of supply, outdated forecasting systems and rigid demand planning systems can make the situation worse (Christopher and Peck 2004).

The demand planning and fulfillment has to be done with a more flexible supply networks with digital enablers. Essential food products and commodities; healthcare and hygiene and OTC nutrition products have seen highest movement in FMCG retail space. Some of the products like sanitizers and hand washes have seen highest intermittent demands. Consumers are regularly looking out for new and updated product offerings which they perceive to be healthier and sustainable. This offers an immense potential for the FMCG players to chalk out the demand of new products and offerings. FMCG stakeholders must not either overestimate or underestimate their demands with a rigid overtone. They must consider more flexible and real-time demand forecasting systems to minimize the idle resources. Optimal sourcing of the goods is thus a result of more resilient demand planning.

Traditional	New Normal
Fast-Moving Consumer Goods	FMCG Classification Higher Demand Fluctuation > Moderate Demand > Low Demand > Negative Demand

Due to the temporary and long-term shift in consumer behavior, FMCG retailers must be very careful in demand forecasting, planning and replenishment. The products classified as non-essential and essentials must be treated on preferential basis depending upon the core competency of retailers and suppliers. All the firms involved in the end-to-end value chain must redesign the processes and realign the operations to either increase or decrease the pace of the stock movement. The FMCG retailers dealing with food and essentials should always be in a lookout for fulfilling a rather unnaturally high stock requirements.

2.4.3 Suppliers Risk Management

With the increase of unexpected panic buying and stocking behavior, key FMCG inventories must be managed on high priority. Due to unending list of store brands, private labels, national brands and global brands of FMCG goods, there is a need for constant supervision of suppliers risks. Firms must identify the company's key suppliers and evaluate their strength and weaknesses to uninterruptedly meet the retail requirements.

One important approach would be to look for a sustainable consumption model, dealing with the consumer goods. Apps, aggregators and other similar service providers can be used to minimize the risk and reinforce the sustainability of an existing model. Every members of a supply chain right from the producers to the consumer will be transformed. In the post pandemic world businesses must strive toward a sustainable model (Sarkis et al. 2020).

Offsetting the supplier's risks can be done with proper management of suppliers on the basis of their core capabilities and proneness toward risk. Evert suppliers inventory, operations and order fulfillment indicators must be captured in real time. Top-level brands with high priority suppliers can accommodate flexible approaches in the execution of production, packaging and order fulfillment. Integrated communications and adaptive contingency plans are to be put in place for any untoward and critical situations arising out of supply chain disruptions.

The role of intermediaries might also be reassessed as it is through them that the smallest producers (micro- and small farmers) may coordinate their actions to produce deeper impact in the provision, and end-consumers can shorten the supply chain by actively involving in production/distribution systems, viz., online delivery and/or community-supported agriculture (Jabbour et al. 2020).

2.4.4 Leveraging Information Technology

There is a need for complete alignment of IT systems in FMCG retail environment. With an outlook of providing 100% efficient safety measures to support the workflow requirements, information technology can be a great savior for the retailers. Right from daily check-ins checkouts and shelf-management to crowd management and monitoring, IT support systems can be a quintessential aspect of safe and healthy value chain. With the provision of remote work and flexible work style, the role for the IT solutions is more relevant than ever. Decentralized supply chains demand a greater role of IT solutions and services.

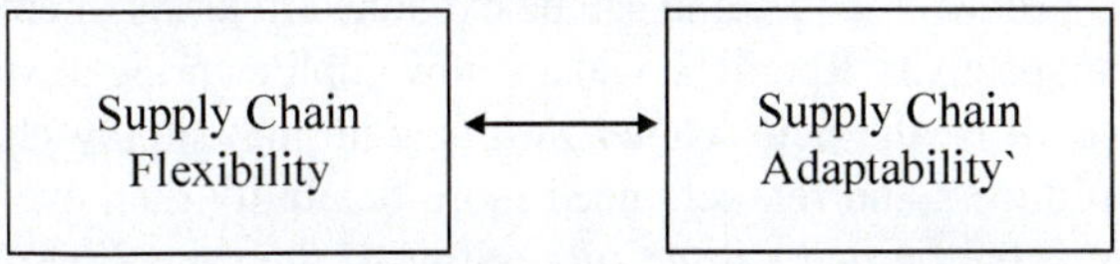

The shift toward non-store retailing formats, viz., online shopping, social media shopping, telephone shopping, WhatsApp shopping etc. can pose serious challenges to the integration ability of IT systems. The order processing and fulfillment without compromising upon data security needs robust networks capacity and high-level testing. At the same time, due to the poor telecom and systems infrastructure minimalistic functions must be enabled. The FMCG retailers and partner firms need to proactively work on the systems requirement, backup support staffs due to absenteeism, critical inventory handling and management to ensure operational efficiency. There is a need to balance both offline and online retail space with IT empowered workforce.

FMCG companies as well as retail companies can be expected to increase the investment in technologies with the capacity of problem-solving. This can be done to materialize the retail plans while making the process more direct, consumer-centric and hygiene-safe. During the COVID-19 pandemic, the timing and sequence of bringing back online facilities at different echelons and nodes of the supply chain had the greatest impact on the supply chain performance (Ivanov 2020).

Better information systems and technologies, encompassing automation systems will be needed by essential supplies service providers, including FMCG sector. There is an increase in demand for a responsive information systems that will need participation from ITES experts in the domain of retailing. In order to quickly bounce back, IT infrastructures linked with supply chain locations, production facilities, materials flow, retail store etc. need real-time synchronization. All members of the supply chain should therefore share information, identify pros and cons and eventually reconfigure supply chain structure in the new setup (Govindan et al. 2020).

2.5 Conclusion

Continuous reconfiguration of supply chain is the new theme in the post COVID world for all the FMCG retailers. The whole world is getting ready to adapt to the new normalcy situation post coronavirus pandemic. The COVID-19 lockdowns across the geographies have catapulted FMCG retail firms toward reinvention of traditional supply chain models. E-commerce, aggregators, large retailers or small kirana stores are restructuring their overall business model, including the supply chain models. Keeping the regular flow of product assortments intact across shelves of various traditional and modern retailers with the right prices and quantity remains a challenge, and at the same time not to compromise with the health and safety.

It will not be an exaggeration to say the effect of COVID saga on consumer spending pattern for FMCG products. This trend can be explained in terms of emergency and essential stock perspectives. Retailers who are now implementing flexible and adaptive supply chain will be the new winner in a new normal supply chain scenario. Today, FMCG producers and retailers need more flexibility than ever in their global and local supply chain process to be successful. At the same time, the positive attitude toward adapting emerging technologies to solve the challenges of FMCG products movement will be a key distinguishing factor.

In order to make supply chains and logistics management more agile and adaptive, there must be a consistent attempt to provide high-quality retail service. It is not an exaggeration to say that supply chains are the life blood of an economy and markets.

References

Choi, T. M. (2020). Innovative'bring-service-near-your-home'operations under corona-virus(COVID-19/SARS-CoV-2) outbreak: Can logistics become the messiah?. *TransportationResearch Part E: Logistics and Transportation Review.*

Christopher, M., & Peck, H. (2004). Building the resilient supply chain. *International Journal ofLogistics Management, 5*(2), 1–13.

Chuu, S.-J. (2011). Interactive group decision-making using a fuzzy linguistic approach for evaluating the flexibility in a supply chain. *European Journal of Operational Research, 213*(1), 279–289.

Govindan, K., Mina, H., & Alavi, B. (2020). A decision support system for demand management inhealthcare supply chains considering the epidemic outbreaks: a case study of coronavirusdisease 2019 (COVID-19). *Transportation Research Part E: Logistics and TransportationReview, 138.*

Hobbs, J. E. (2020). Food supply chains during the COVID-19 pandemic. *Canadian Journal of Agricultural Economics.*

Huber, J., Gossmann, A., & Stuckenschmidt, H. (2017). Cluster-based hierarchical demand forecasting for per-ishable goods. *Expert Systems with Applications, 76,* 140–151.

Ivanov, D. (2020). Predicting the impacts of epidemic outbreaks on global supply chains: A simulation-based analysis on the corona-virus outbreak (COVID-19/SARS-CoV-2) case. *Transportation Research Part E: Logistics and Transportation Review.*

Jabbour, C. J. C., et. al. (2020) Sustainability of supply chains inthe wake of the coronavirus(COVID-19/SARS-CoV-2)pandemic: lessons and trends. *Modern Supply Chain ResearchandApplications.*

Liu, C. L., & Lee, M. Y. (2018). Integration, supply chain resilience, and service performance in third-party logistics providers. *The International Journal of Logistics Management.*

Mihajlović, S. L., & Trajković, S. (2020). The importance of logistics and supply chains for pandemia conditions. *Journal of Process Management. New Technologies, 8*(2), 53–59.

Sarkis, J., Cohen, M. J., Dewick, P., & Schroder, P. (2020). A brave new world: lessons from theCOVID-19 pandemic for transitioning to sustainable supply and production. *Resources, Conservation, and Recycling.*

Scheibe, K. P., & Blackhurst, J. (2017). Supply chain disruption propagation: A systemic risk and normal accident theory perspective. *International Journal of Production Research,* 1–17.

Yu, H., Sun, X., Solvang, W. D., & Zhao, X. (2020). Reverse Logis-tics network design for effective management of medical waste in epidemic outbreaks: insights from the coronavirus disease 2019 (COVID-19) outbreak in Wuhan (China). *International Journal of Environmental Research and Public Health, 17*(5), 1770.

Yu, K. D. S., & Aviso, K. B. (2020). Modelling the economic impact and ripple effects of disease outbreaks. *Process Integration and Optimization for Sustainability,* 1–4.

Chapter 3
Challenges Faced and Preparedness of Agriculture Supply Chain During COVID-19

Nitish Maan, Vijaya Kumar Manupati, Maciel M. Queiroz, and Biswajita Mohanty

Abstract The coronavirus (COVID-19) pandemic created catastrophic failure on global supply chains, particularly its impact on the food supply chain left havoc across the globe. The objective of this study is to understand the challenges imposed by the pandemic to the agricultural supply chain in the context of the Indian subcontinent and also analyse the variations in the arrivals and costs of the products. As a first step, the conventional flow of perishable food supply chain products from farmers to end customers has been described. Thereafter, the most widely used perishable product (tomato) has been chosen and its availability across the Indian regions has been identified. Consequently, the zones (red, green, and orange) under which these regions are mapped are identified. Later, with several analyses, the effect of zones and its impact on product arrival and its cost are discussed in detail. Finally, several research directions and the challenges to overcome the agricultural supply chain have been discussed.

Keywords Agricultural supply chain · Pandemic · Conventional supply chain · Zones

N. Maan · V. K. Manupati (✉)
Department of Mechanical Engineering, National Institute of Technology Warangal, Warangal, India
e-mail: manupati.vijay@nitw.ac.in

N. Maan
e-mail: maan_193513@student.nitw.ac.in

M. M. Queiroz
Postgraduate Program in Business Administration, Paulista University–UNIP, São Paulo 04026-002, Brazil
e-mail: maciel.queiroz@docente.unip.br

B. Mohanty
Vellore Institute of Technology Vellore, Vellore, Tamil Nadu, India
e-mail: mohanty.biswajita@vit.ac.in

A. R. Sakthivel et al. (eds.), *Managing Supply Chain Risk and Disruptions: Post COVID-19*, Management and Industrial Engineering,
https://doi.org/10.1007/978-3-030-72575-4_3

3.1 Introduction

The agricultural products both perishable and non-perishable together shaped the food supply chain (Nahmias 2011), where the perishable products are those that spoil most quickly and have a short life span; on the other hand, the non-perishable products are those not constrained by time. By their nature, the perishable products demand a robust supply chain (SC) that ensures fast and efficient distribution of products, and the non-perishable products demand resilient SC due to its strenuous procurement and distribution strategies. Considering the high population density of India, which constitutes around one-sixth of the world population, the complexity of the SC is further compounded.

The Indian agricultural SC mainly consists of three stages: it starts with the farmers selling their products in the regional mandis, from where the wholesalers distribute it to the retailers, and from the retailers, it reaches to the end customers. The distribution of products may seem simple in overview but while considering the time constraint in perishable and storage constraint in non-perishable products along with the transportation across the state border, it gets quite complex. India stands at 102 among 117 countries in the Global Hunger Index 2019, and hence it is important to measure the impact on the agriculture SC and recognize the challenges posed by the COVID times. Further, COVID-19 leaves challenges across the streams in the SC, irrespective of various flexibilities that exist at each stage, particularly the food SC is more vulnerable due to the disruptions at upstream, i.e., harvesting and distribution stages. Moreover, the perishable SC is more volatile than the non-perishable SC as it needs regular supply to the mandis, and factors like the fear of spread, lack of labour, and sealed state borders have resulted in a decline in the supply.

Due to nationwide lockdown, the agricultural SC has affected all streams simultaneously never before, starting from farmer's end, the harvesting of the rabi crops was at stake due to lack of labour, and harvesting machinery thus impacting the upstream. The distribution from the wholesalers got affected and the products could not reach the retailers due to halt and clearance; it takes time in initiating the protocols of the transportation system by respective state governments. Finally, the product is delivered to the end customer and the mode of distribution in case of perishable products consists of city mandis, weekly markets, and street vendors. However, the major distribution is done by the weekly markets and city mandis, whereas the ratio of distribution by the street vendors is low. The weekly markets were suspended completely to curb the community spread and the main markets across many cities were either closed or shifted due to reporting of COVID-19 cases although it was later on compensated by increasing the street vendor capacity. In the case of the non-perishable products, the time buffer gave some advantage and the previous stock was also utilized to fulfil the demand.

The literature is reviewed based on the perishable product SC, disruptions, and resilience of the SC during the epidemic, and the effects of COVID-19 on the SCs of the world in terms of the economy and food security as well. In December 2019, a series of pneumonia cases emerged in China, later identified as the novel coronavirus and named severe acute respiratory syndrome coronavirus 2 (SARS-CoV-2), also known as the disease name COVID-19 (Tan et al. 2020), and with the official announcement of a pandemic, the study of internet searches and streaming depicted that demand for food and adaptation of healthy food habits grew (Laguna et al. 2020). The demand for food products was meant to increase but the pandemic brought lockdowns that disrupted the SC at various levels. The major research developed for a pandemic is in the area of controlling pandemic with medical resource allocation and vaccine distribution models (Chick et al. 2008; Mamani et al. 2013; Liu and Zhang 2016; Büyüktahtakın et al. 2018). A recent study by Queiroz et al. (2020) reviewed the impact of the epidemic on commercial SCs and their findings revealed that resource allocation and distribution gain its importance. Golan et al. (2020) identified lack of resilience in SCs created cascading failure under disruptions where the current pandemic shows its proof. Hence, there is a need to review SC resilience and its modelling on networks, including transportation, command, and control. The loss of agri-business revenue and rise in food insecurity will be the results of the disruption in the SC as discussed by the FAO (Food and Agriculture Organization 2020) and impact on the farmers of Haryana (Ceballos et al. 2020) with reports saying they are unable to afford enough food during the lockdown shows the food insecurity created due to loss in revenue. In this chapter, the possibilities to overcome the challenges that occurred due to the pandemic are explored and also find the possible policies taken by different state governments across the Indian nation that has brought provisions for buying the products at minimum support price as well as providing financial assistance.

3.2 Study Area Map of Mandis and Their COVID-19 Zones

In this section, the conventional flow of perishable food SC in the Indian subcontinent is discussed in detail. The flow of perishable products can be broadly categorized into three categories, which start with procurement in which the farmer sells the product in the nearby mandis called Mandi A. From Mandi A, the wholesaler who buys the product takes care of transportation to move to its regional mandis. The wholesaler can be from within the state as well as from the other state accomplishing the inter-state and intra-state transportation, as shown in Fig. 3.1. After the product arrives it has been moved to the nearby mandis (Mandi B and Mandi C). Now the third stage of the flow starts with the distribution of the product to the end consumer. The local retailer and vendors buy the product from the wholesaler at the local mandis (Mandi B and Mandi C) and distribute the product to the end customers. This flow of products across perishable SC during the normal times can be called the conventional SC, as depicted in Fig. 3.1.

The above-shown SC depicts the flow of a single product, but in reality, there is N number of products flowing into the regional mandis of inter-state and intra-state via wholesalers, and a single retailer will buy M different products in small

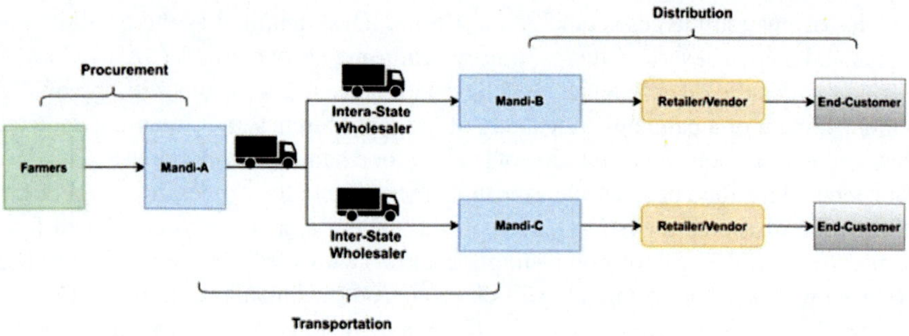

Fig. 3.1 The conventional flow of perishable product

quantities from different retailers and distribute them to the end consumer. In this study, tomato as a perishable product is considered due to its wider use and less shelf life. Table 3.1 describes the number of states across India the tomato product is available and their districts with existing mandis. The collected mandis location for considered perishable product and their respective zones is depicted in Fig. 3.2.

3.3 Analysis of Perishable Products in Different Mandis During COVID-19

In this section, different parameters of perishable products SC and its behaviour and effects on COVID-19 have been analysed. However, the nature and behaviour of SC for other perishable products is almost the same; in this study as the perishable product tomato is considered. To understand the effects of COVID-19 on complex perishable SC, the analysis with two variables, i.e., "Arrival Quantity" and "Model Cost" has been considered. The survey has been conducted with tomato abundant/harvesting states, districts, and mandis available in India, and its arrival (in quintals) and price (Rs/quintal) data were acquired for the month of January to June for the years 2017 to 2020 from the National Horticultural Research and Development Foundation (NHRDF) portal. Microsoft Excel and Python software have been used to perform data cleaning, and then the mean of arrival quantity and mean price for the past years were calculated with which the percentage change in arrival (PCAR) quantity and percentage change in price (PCP) with the 2020 data are established. Based on the available data, 37 mandis were selected across different states as shown in Table 3.2. For example, in the market of Delhi state, the arrival (in quintals) for the month of May from 2017 to 2019 be a_1, a_2, and a_3, while for 2020 be a_4. The mean for a_1, a_2, and a_3 is denoted by a_{mean}. Then the PCAR is given by

$$PCAR = \frac{a_4 - a_{mean}}{a_{mean}} * 100$$

Table 3.1 Considered state, district and mandi

S. No.	State	District	Mandi
1	Andhra Pradesh	Chittoor	Mulakalacheruvu
2	Chhattisgarh	Bilaspur	Thipra
3	Delhi	Delhi	Delhi
4	Gujarat	Ahmedabad, Surat, Rajkot, Bhavnagar	Ahmedabad, Surat, Rajkot, Bhavnagar
5	Himachal Pradesh	Kullu, Mandi, Solan	Kullu, Mandi, Solan
6	Haryana	Faridabad, Gurugram	Faridabad, Gurugram
7	Jammu & Kashmir	Jammu	Jammu
8	Kerala	Thiruvananthapuram	Palayam
9	Karnataka	Mysore, Kolar, Chickaballapur, Bengaluru (Binny Mills)	Mysore, Kolar, Srinivaspur(Kolar), Binny Mills, Chickballapur, Chintamani(Chickballapur)
10	Madhya Pradesh	Mandsour	Mandsour
11	Maharashtra	Aurangabad, Pune Chandrapur, Kolhapur, Mumbai, Nagpur, Solapur	Aurangabad, Chakan (Pune), Pune, Chandrapur, Kolhapur, Mumbai, Nagpur, Solapur
12	Odisha	Nayagarh, Bhubaneswar, Ganjam, Betnoi (Mayurbhanj), Bargarh	Bahadaholla (Nayagarh), Bhubaneshwar, Hinjlicut (Ganjam), Betnoi (Ganjam), Bargarh
13	Punjab	Amritsar, Fazilka (Abohar), Bhatinda, Firozpur, Jalandhar	Amritsar, Abohar(Fazilka), Bhatinda, Firozpur, Jalandhar
14	Rajasthan	Ajmer, Alwar, Jaipur, Jodhpur, Sriganganagar, Udaipur	Ajmer, Alwar, Jaipur, Jodhpur, Sriganganagar, Udaipur
15	Telangana	Hyderabad	Bowenpally, Gudimalkapur (District-Hyderabad)
16	Uttar Pradesh	Agra, Prayagraj, Ballia, Barely, Deoria, Etah, Etawah, Faizabad, Saharanpur, Sitapur, Varanasi, Aligarh	Agra, Allahabad (Prayagraj), Ballia, Barely, Etah, Etawah, Faizabad, Saharanpur, Sitapur, Varanasi, Aligarh
17	West Bengal	West Bardaman (Asansol), Purba Bardhaman (Burdwan)	Asansol, Burdwan

The PCAR of Delhi for May month is −51.74%, which signifies that the arrival for that particular month has decreased by 51.74% in 2020 as compared to the mean of the arrival during the previous years.

Similarly, the price (Rs/quintal) for the month of May from 2017 to 2019 be p_1, p_2, p_3, while for 2020 be p_4. The mean for p_1, p_2, p_3 is denoted by $p_{\text{mean.}}$ Then the PCP is given by

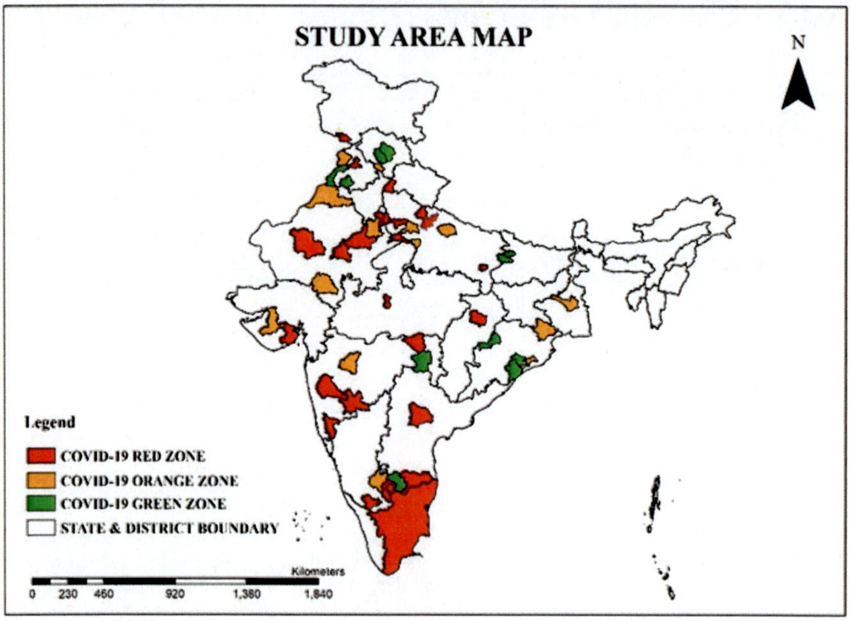

Fig. 3.2 Study map having mandis and zones

$$PCP = \frac{p_4 - p_{mean}}{p_{mean}} * 100$$

The PCP for the same month is −18.80% showing the fall in price for May 2020 by 18.80% as compared with the mean of previous years' model price. Later on, the data were plotted in graphs and maps, and the insight into the variations was studied.

3.4 Findings

The PCAR and PCP of different mandis were calculated from January to June by the method as discussed in Sect. 3.3 for various states across India. With the help of a histogram, the PCAR and PCA for particular mandis are shown in Fig. 3.3. The red zone mandis, Mumbai (Maharashtra), Delhi, Faridabad (Haryana), Agra (Uttar Pradesh), and Bhavnagar (Gujarat), show a trend of negative PCAR throughout, i.e., the arrival has decreased as compared to the previous year's highlighting the effect COVID-19 on the SC of perishable products. Whereas from the developed histogram, Mumbai from Maharashtra state follows downward trend in PCAR that can be seen as −93.26% for the month of May. Looking at the PCP for Mumbai, the graph depicts a negative trend in that also, but for the Aurangabad mandi of Maharashtra state, the PCP shows a negative shift as well, maintaining a fall of around 40% throughout and

Table 3.2 Percentage change in arrival quantity and modal price in different mandis in India

S.No.	State	Market	% Change of arrival						% Change of price					
			Jan	Feb	Mar	Apr	May	June	Jan	Feb	Mar	Apr	May	June
1	Andhra Pradesh	Mulakalacheruvu	−22.06	−43.98	−8.40	120.60	0.77	−22.69	27.44	21.39	−5.44	−22.32	−64.38	−46.39
2	Chhattisgarh	Thipra	−37.74	−71.64	−75.18	141.02	39.78	−1.48	18.65	5.07	−10.37	−6.17	−35.55	−8.85
3	Delhi	Delhi	−26.98	−85.21	NA	−5.19	−51.74	−58.12	90.18	17.98	NA	10.32	−18.85	−30.53
4	Gujarat	Ahmedabad	−13.29	−3.75	22.94	−67.55	NA	−70.32	74.32	−16.52	−51.68	−34.58	NA	36.13
		Bhavnagar	−61.65	−69.94	−58.48	−52.05	−60.44	−44.41	11.43	8.07	−14.64	−20.93	−14.44	27.77
		Rajkot	45.81	3.05	40.36	−7.19	37.78	234.66	24.99	−17.43	−50.00	−5.62	−21.29	−19.42
		Surat	6.15	−39.58	−37.68	−53.90	−81.82	−59.95	0.17	−54.48	−47.66	2.65	−44.91	1.24
5	Himachal Pradesh	Kullu	55.28	−8.67	−35.88	−4.17	27.57	32.83	23.83	−15.75	7.15	14.53	−3.16	−6.71
		Mandi	−69.33	−93.48	−86.54	−54.46	−70.77	−15.82	22.94	−16.57	12.18	24.99	−26.36	−21.69
		Solan	15.72	−34.67	−19.82	134.90	−55.91	−98.92	29.49	−9.80	−9.34	22.72	−19.32	−54.44
6	Haryana	Faridabad	12.53	47.56	14.16	50.23	18.54	−66.56	45.88	14.06	−5.89	35.95	−47.05	−50.46
		Gurugram	−66.84	−43.97	−41.97	79.96	155.25	−4.79	−18.56	−15.07	−11.80	107.73	−44.00	20.00
7	Jammu & Kashmir	Jammu	NA	−78.69	−35.70	462.04	15.38	−7.51	NA	−0.58	19.72	27.77	19.55	28.30
8	Kerala	Palayam	0.00	−33.33	−34.50	−41.46	−68.91	−61.25	−3.79	1.45	−26.96	−36.98	−48.23	−23.67
9	Karnataka	Binny Mills	−10.21	−31.18	−41.63	248.60	88.56	−26.80	29.78	−7.72	−32.89	−47.40	−53.86	−11.61
		Chickaballapur	−14.91	−76.16	−97.87	46.51	−29.14	57.59	54.74	−31.42	−29.69	−53.32	−62.48	0.23
		Chintamani	−56.96	−85.77	−95.89	10.79	−30.35	−89.49	−7.31	−57.16	−55.58	−54.94	−68.90	−41.05
10	Madhya Pradesh	Mandsour	−10.80	−57.40	−74.47	NA	−67.77	−76.67	1.68	32.12	20.91	NA	−3.97	7.21

(continued)

Table 3.2 (continued)

S.No.	State	Market	% Change of arrival						% Change of price					
			Jan	Feb	Mar	Apr	May	June	Jan	Feb	Mar	Apr	May	June
11	Maharashtra	Aurangabad	−32.53	0.42	−69.42	−47.51	−43.72	−42.22	37.94	−35.14	−41.95	−41.83	−46.74	−16.82
		Mumbai	10.20	−53.36	−55.02	−85.28	−93.26	−81.47	16.78	−5.34	−15.90	−2.71	−17.72	−1.91
		Nagpur	4.88	−56.36	−87.88	−90.20	−89.36	−52.87	15.08	−18.29	−14.19	−32.53	−32.45	44.74
12	Orissa	Bahadajhola	−30.49	−89.77	−98.08	−99.38	−94.36	−83.89	6.18	−17.46	−14.67	−26.61	−37.30	−39.46
		Bargarh	NA	−12.26	−16.99	−33.98	−72.60	−52.15	62.70	−46.49	−12.36	197.17	71.57	−46.34
13	Punjab	Abohar	−44.51	−20.18	−56.09	−61.22	−95.65	−52.70	46.07	−21.74	−21.58	36.56	25.26	−45.85
		Amritsar	109.71	35.33	−4.28	4.93	−22.74	−45.67	0.36	−26.13	−27.12	17.57	−46.17	−7.45
		Barnala	59.54	−15.38	26.71	−92.71	−41.42	2.49	10.32	−24.91	−21.77	38.77	−31.42	−36.11
		Firozpur	401.23	107.53	−11.63	32.13	0.96	−28.51	38.50	−6.55	−21.69	47.56	−5.98	4.17
14	Rajasthan	Ajmer	127.32	−34.22	−20.70	−27.98	−83.35	−76.65	44.11	−3.87	−20.46	46.78	−21.27	−37.51
		Alwar	−48.75	−79.78	−40.04	−24.81	−67.66	−41.14	68.22	22.15	1.19	−7.87	−42.88	28.50
15	Telangana	Gudimalkapur	−36.57	NA	−50.59	25.85	−32.79	−18.31	61.32	NA	−28.37	−56.98	−54.54	−0.42
16	Uttar Pradesh	Agra	−5.53	−18.94	−57.93	−61.66	−47.85	−36.24	58.56	34.39	−10.80	31.10	−13.32	−17.62
		Gaziabad	48.53	−45.48	−67.00	−22.03	−27.95	−64.47	33.01	25.77	−15.03	50.96	13.09	−14.14
		Kanpur	−55.57	−74.63	−75.95	−55.50	−75.37	−70.00	74.48	72.74	7.02	56.41	12.96	−12.48
		Saharanpur	−16.04	−62.37	−61.71	2.71	−46.43	−46.42	60.39	43.80	9.35	66.30	49.12	−4.78
		Lucknow	−17.69	−47.51	−34.07	−29.10	−10.40	−4.17	67.57	76.56	13.26	25.03	−12.85	−14.77
17	West Bengal	Asansol	−67.29	−14.71	−73.90	1.59	−57.96	−77.01	40.07	48.66	−1.76	−24.71	−35.42	−1.94

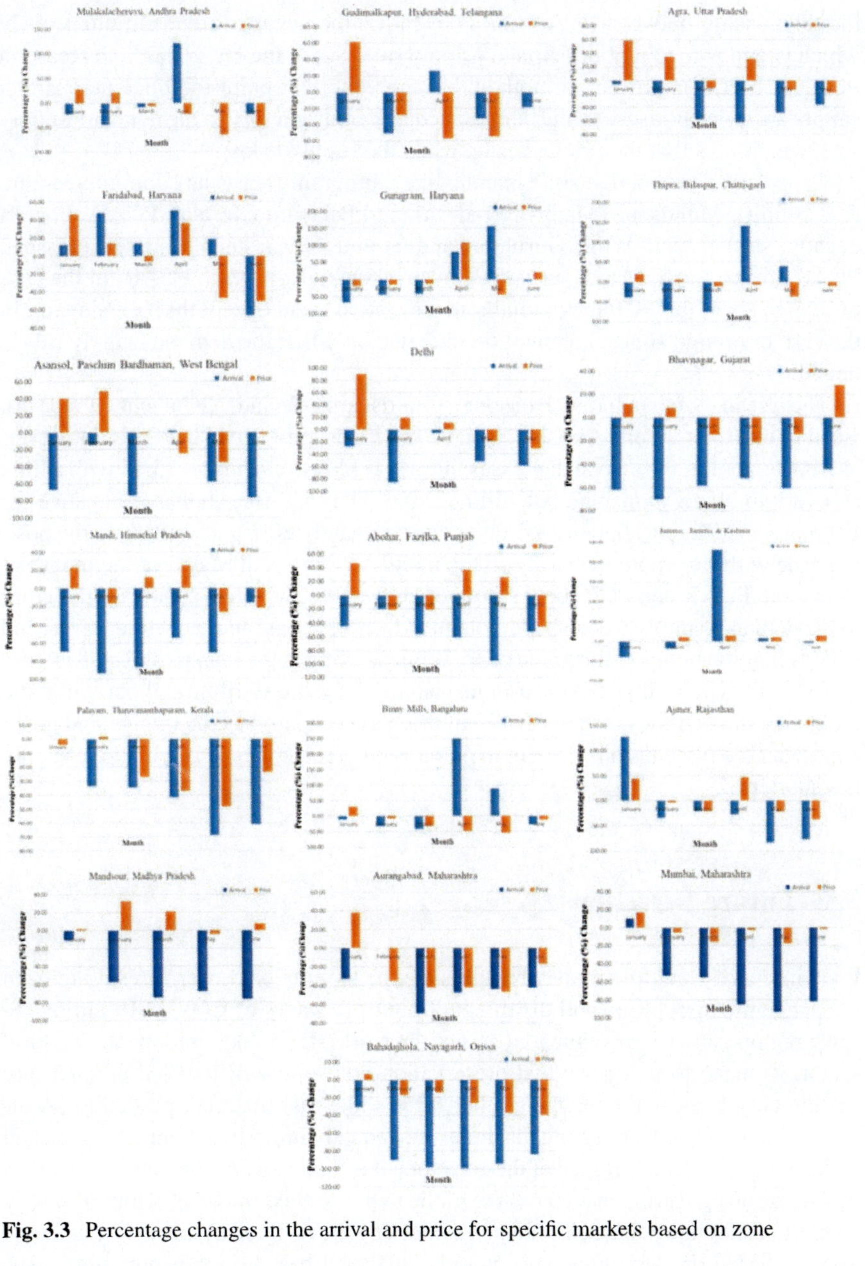

Fig. 3.3 Percentage changes in the arrival and price for specific markets based on zone

the relationship between PCAR and PCP shows the negative trend simultaneously which is opposite to that of demand–supply rule. So, in the red zones, a decrease in PCAR and PCP is present throughout but the important point to notice here is that supply and price both are falling simultaneously contradictory to the demand–supply rule, which says that the price goes up when the supply is less.

Considering the orange zone mandis like Gurugram (Haryana), Jammu (Jammu & Kashmir), Mandsour (Madhya Pradesh), and Palayam (Kerala), PCAR shows a negative shift as well. While Gurugram and Jammu show a steep positive increase in PCAR during April, Mandsour and Palayam show a decrease in PCAR. In the case of orange zone, the PCP is not totally in a negative trend except that of Palayam. In the case of orange zones, it cannot be said evidently that the demand–supply rule is defied.

The green zone mandis, Bilaspur (Chattisgarh), Mandi (Himachal Pradesh), Bhadajhola (Orissa), and Abohar (Punjab), show a negative trend in PCAR. However, at Abohar mandi, the arrival decreases linearly till May reaching −95.65 with a little recovery to −52.7% in June but still negative. The PCP for Abohar is negative for February and March, showing a drop of approximately 20% and shifting to the positive side with rise more than 20% in the month of April and May. For the month of June, both PCAR and PCP show a drop of above 40%, which is again contradictory to the demand–supply rule. Bhadajhola mandi shows consistent negative PCAR and PCP even if it belongs to the green zone. While looking at the mandis and considering zones, it is evident that the drop-in arrival can be witnessed throughout while the red zones showed the consistent drop in price as compared to the orange and green zones, but the price drop is present in orange and green zones also even if the supply has shortened.

3.5 Future Directions

Food insecurity and loss in agri-business are the two major concerns resulting from the pandemic disruption, and the threat of multiple waves of COVID-19 hitting the same region can also be witnessed across the world. If we look into the agricultural sector, we need to recognise that preservation and supply of finished products are not the only goal, but identifying other factors like raw material, pesticides, availability of labour, and harvesting machinery are equally important. For identifying and understanding the importance of these factors, it is vital to study the policies taken up by the various governments across the globe in the context of the agricultural supply chains to tackle the pandemic and then evaluating the resilience and cost parameters. The SC based on the policies needs to be analysed based on transportation costs, inventory, and ability to serve rural as well as urban populations. In India, 95% of the purchased goods are sold by the private sector and only 5% through public distribution system (PDS) (Reardon et al. 2020), so the SC cannot be fully controlled by the government and the policies need to be private sector-oriented. These policies can further be mixed and evolved to prepare not just for the ongoing crises but any future threats as well.

3.6 Conclusion

The COVID-19 pandemic has stalled every SC globally causing disruptions, and an analysis here depicts that there is a disruption in the supply of perishable products to the markets or mandis as well. The reasons for the disruption in the flow of products to the mandis can be lack of labourer workers, implementation of nationwide lockdown, fear of COVID spread among the public, and creation of containment zones to curb the spread which also disrupts the flow of end-consumer to the local mandis. If we look at the SC of perishable products as shown in Fig. 3.1 with the three stages of procurement, transportation, and distribution, as the supply of product declined to the mandis, we can speculate the disruption in the procurement stage, i.e., farmers are not able to deliver the product to the mandis and it can be inferred that the distribution of product to the end-consumer is also declining because food is an essential product and the decrease in demand is not certain from the consumer but can be due to inaccessibility of the end-consumer to reach the product which can lead to a decline in the price of the product as well. Thus, if the SC of perishable products is disrupted, then to save the value of the product, it needs to be preserved for which the demand for cold storage facilities, refrigerated carriers, proper packing, and handling and real-time tracing comes into play.

References

Büyüktahtakın, E., des-Bordes, E., & Kıbış, E. Y. (2018). A new epidemics–logistics model: Insights into controlling the Ebola virus disease in West Africa. *European Journal of Operational Research, 265*(3), 1046–1063. https://doi.org/10.1016/j.ejor.2017.08.037.

Ceballos, F., Samyuktha, K., Berber, K. (2020). Impacts of a national lockdown on small-holder farmers' income and food security: empirical evidence from two states in india. *World Development, 136*, 105069. https://doi.org/10.1016/j.worlddev.2020.105069.

Chick, S. E., Mamani, H., & Simchi-Levi, D. (2008). Supply chain coordination and influenza vaccination. *Operations Research, 56*(6), 1493–1506. https://doi.org/10.1287/opre.1080.0527.

Food and Agriculture Organization. (2020). Addressing the Impacts of COVID-19 in Food Crises|April–December 2020: FAO's Component of the Global COVID-19 Humanitarian Response Plan. https://doi.org/10.4060/ca8497en.

Golan, M. S., Jernegan, L. H., & Linkov, I. (2020). Trends and applications of resilience analytics in supply chain modeling: Systematic literature review in the context of the COVID-19 pandemic. *Environment Systems & Decisions*, 1.

Laguna, L., Fiszman, S., Puerta, P., Chaya, C., & Tárrega, A. (2020). The impact of COVID-19 lockdown on food priorities Results from a preliminary study using social media and an online survey with Spanish. *Consumers Food Quality and Preference*. https://doi.org/10.1016/j.foodqual.2020.104028.

Liu, M., & Zhang, D. (2016). A dynamic logistics model for medical resources allocation in an epidemic control with demand forecast updating. *Journal of the Operational Research Society, 67*(6), 841–852. https://doi.org/10.1057/jors.2015.105.

Mamani, H., Chick, S. E., & Simchi-Levi, D. (2013). A game-theoretic model of international influenza vaccination coordination. *Management Science, 59*(7), 1650–1670. https://doi.org/10.1287/mnsc.1120.1661.

Nahmias, S. (2011). Perishable inventory systems. *International Series in Operations Research and Management Science, 160*. Springer.

Queiroz, M. M., Ivanov, D., Dolgui, A., & Wamba, S. F. (2020). Impacts of epidemic outbreaks on supply chains: Mapping a research agenda amid the COVID-19 pandemic through a structured literature review. *Annals of Operations Research* 1–38.

Reardon, T., Mishra, A., Nuthalapati, C., Bellemare, M., & Zilberman, D. (2020). Covid-19's disruption of India's transformed food supply chains. *Economic and Political Weekly, 55,* 18–22.

Tan, W., Zhao, X., Ma, X., Wang, W., Niu, P., Xu, W., et al. (2020). A novel coronavirus genome identified in a cluster of pneumonia cases—Wuhan, China 2019–2020. *China CDC Weekly, 2*(4), 61–62. https://doi.org/10.46234/ccdcw2020.017.

Chapter 4
Challenges Faced and Preparedness of Humanitarian Supply Chain During COVID-19

R. Vezhavendan

Abstract Today, we are faced with a pandemic which has already infected several millions of people worldwide and the virus is out on a spree to infect more and more people. This scenario is likely to continue until a vaccine is found and administered to a majority of the population. The human race had not anticipated such a disruption of supply chains on such a massive scale. All the transportation and movement of goods had come to a standstill for a varied period of time in various countries. This resulted in disruptions and shortfalls in various parts of the world. The demand for several products fell, while demand for several products increased conversely. A humanitarian supply chain could be described as relief material supplies reaching the right people at the time of need. This is required to save several precious lives from being lost due to lack of access to products that are in need. This book chapter comprises discussion on the various challenges faced by relief works that need to be executed and how we need to be prepared for any eventualities that may arise at any point of time. This chapter is intended to provide insights into various logistic issues and preparedness levels of mankind to face the challenges posed by COVID-19.

Keywords Humanitarian supply chain · Goods · Challenges · Emergency preparedness

4.1 Introduction

Humanitarian supply chain (HSC) is the term coined to describe the flow and supply of relief materials to reach the right people at the right time, especially when disasters strike at unexpected times. Today, we are faced with a pandemic due to the spread of the corona virus (COVID-19) which is considered to have originated from the wet market of Wuhan, China. As the virus has spread to every nook and corner of the world, it has caused unprecedented disruptions and damage to human lives

R. Vezhavendan (✉)
Department of Manufacturing Engineering, School of Mechanical Engineering, Vellore Institute of Technology, Vellore 632014, Tamil Nadu, India
e-mail: vezhavendhan.r@vit.ac.in

A. R. Sakthivel et al. (eds.), *Managing Supply Chain Risk and Disruptions: Post COVID-19*, Management and Industrial Engineering,
https://doi.org/10.1007/978-3-030-72575-4_4

over the globe along with both direct and indirect damage to several products. The virus outbreak has caused a huge unexpected humanitarian crisis and resulted in unprecedented challenges to human lives and health across the globe, and the global economy as a whole has taken a hit. There does not seem to be any respite for the outbreak, and on the other hand, mankind is faced with several severe shortages, especially products related to food and health. We can say that presently we are faced with a "supply chain shock" due to the disruptions worldwide. The world was not anticipating even anything close to such a thing and was completely unprepared for such a massive outbreak on a global scale. This crippled several activities along the lives of the whole mankind. Nothing of this sort was expected and there was absolutely no preparation. Had the world expected such a massive outbreak and anticipated things to go haywire over the COVID-19 virus, things could have been completely different. Moreover, the whole of humankind did not expect the COVID-19 pandemic to be lasting for several months which could be extending for a much longer time.

4.2 The Humanitarian Supply Chain

The impact of COVID-19 can be expected to be short-lived, but the effect it may have in the long term at this point of time is highly unpredictable. We can expect and experience unprecedented stress on the supply chains related to food due to the distress faced due to shortage of farm labour, production shortfalls, procurement shortfalls, and problems faced by the transport and logistics sector coupled with a shift in the demands for products. The moves that many governments of several countries have taken, especially the lockdowns to contain the spread of the COVID-19, had resulted in stoppage of production of several essential commodities that would be required for humanitarian aid. Further, facts have pointed out that the uncertainties that have been caused by such moves, which is known as the "Whiplash effect", refer to the fluctuations seen in the inventory due to the fluctuations in demand. As the demand increases, the inefficiencies in the supply chain can be felt. This is due to panic-buying attitude of customers, wherein customers fear unavailability of certain products (such as hand sanitizers and masks) and end up buying and stocking more than what is needed to stay comfortable later. The supply chains based on humanitarian relief have to provide the correct quantity of products at the right time to the right locations, so as to help relief workers to save lives and abet suffering of humans which are constrained by limited financial backing (Beamon and Balcik 2008). It is believed that pre-positioned warehouses located at strategic points would be a necessity and also be of great value and help to see to it that the supplies needed are available to the needy when they are required and for the facilitation of faster responses (Balcik et al. 2010).

A typical humanitarian supply chain environment is generally complex and challenging (Seyed Hamid et al. 2020). It comprises people at the disaster struck site (as

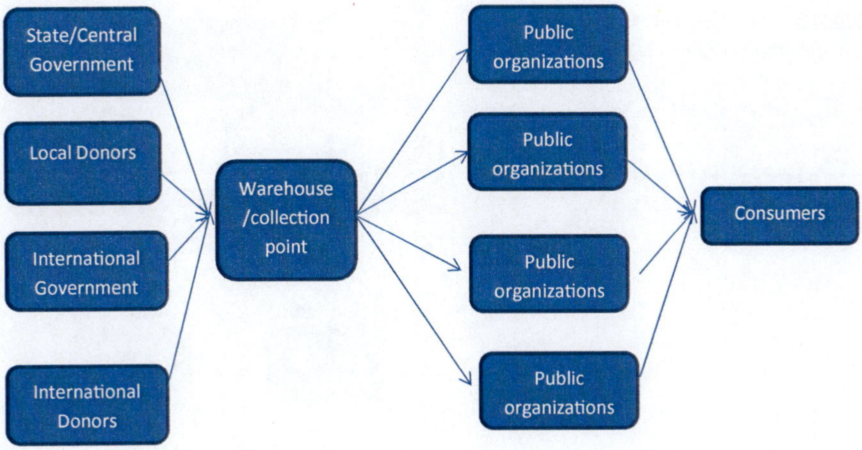

Fig. 4.1 A typical humanitarian supply chain flow

the consumer) to whom all humanitarian aid needs to be supplied. Unlike commercial supply chains, the suppliers are the government of the state or country, local donors, international governments and international donors. The products sent by these donors reach the nearest port or warehouse or collection point where it is needed and the distribution begins through the public organizations, private organizations, community-based organizations and individual volunteers. These act as middle men in the supply chain to reach the consumers, with no monetary profit in mind. Figure 4.1 is a representation of a typical humanitarian supply chain.

There are five important components of the humanitarian supply chain, namely man, money material, machine and information. All the five components need to be available at a time at the required places for the humanitarian supply chain to be up and effective. Men in the form of volunteers and government/public organizations need to be deployed who need money to get required material from various sources such as government of the state or centre or international governments donors who may be from within the state or country or even international donors and these materials required to be processed by machines such as vehicles that can travel on land, water and air and information for and to all these results in effective distribution of humanitarian aid. The right kind of men need to be there at the right place and at the right time to effectively use the material received from various sources.

Money would help in mobilizing the right kind of material and help in putting in place the right machines to effectively execute the work for which all these resources have been put in place. Figure 4.2 represents the various components that need to come together for the humanitarian supply chain to be effective.

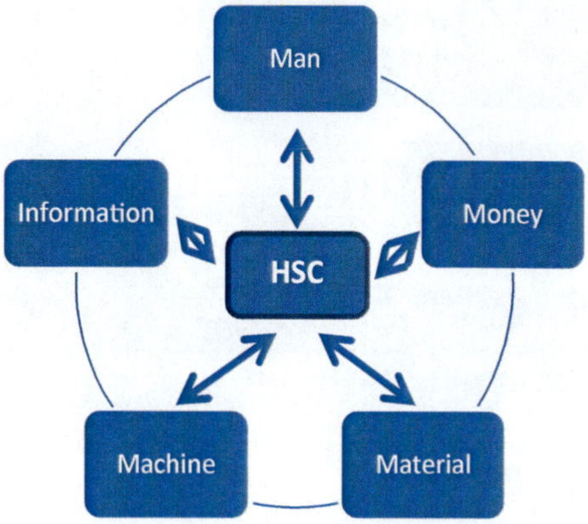

Fig. 4.2 The components of a humanitarian supply chain

4.3 The Role Played by Humanitarian Supply Chains

The COVID-19 pandemic has disrupted the lives of several millions of people around the globe, especially in the developing countries like India, and several other Asian countries along with their public health systems, by bringing the livelihood and economies all over the world to a grinding halt. While countries with strong economies have been able to manage the onslaught, the countries with weak economies comprising population depending on day-to-day earnings for their bread and butter have been the worst hit. Humanitarian supply chains play a vital role when such disaster strikes. Words are no match to express the huge solace that humanitarian supply chains provide during bad times. Their contribution in the mitigation of the negative effects of the disasters with respect to human life loss and economic losses are incomparable. Aids on humanitarian grounds are rushed to the areas with vulnerable population, especially to the population facing food crisis and the population affected by health crisis. Responses to disasters are aimed at offering help to maximum number of victims possible and as quickly as possible without taking much consideration of the socio-economic context (Kunz and Gold 2017).

4.4 Commercial Supply Chain Versus Humanitarian Supply Chain

According to www.investopedia.com, a commercial supply chain is referred to as simply supply chain which is a network between suppliers and customers to supply customer-desired products to the end users, also known as final buyer. This network is a consortium of varied elements comprising activities, men, information, resources, organizations etc. The supply chain is also a representation of all the steps it takes for the product to reach the customer from the supplier from its original state. It comprises a series of steps and activities that are involved before the customer receives the products or services, which may involve movement of goods and transformation of raw materials into finished products, followed by transportation of the finished products and distribution of the same to the end user/consumer at the least possible price and make highest profits possible. The suppliers (vendors), warehouses, transporters, distributors and retailers form the entities who are involved in the supply chain. The supply chain comprises functions that start right from order receiving to supplying the product to the customer. The functions of a supply chain comprise activities starting from product development, marketing, production, operations, distribution of the products, finance and provision of service to customers.

The International Federation of Red Cross (IFRC) defines humanitarian supply chain as procuring and supplying supplies and providing services that have been requested at the place where and when they are needed such that the best value for money is ensured. The requirements immediately after the disaster are generally food, water, temporary shelter and medicine, which are necessary for survival. Humanitarian supply chains perform the task of linking the suppliers to the customers which is the fulfilment of demand by supply. Ergun et al. (2009) have mentioned that generally a typical humanitarian supply chain comprises governments and non-governmental organizations (NGOs). The government has control over both political and economic conditions and decisions which have a direct effect on the process of supply chain activities. There are several teething problems associated with the coordination and management of supply chains related to disaster management activities because there are several uncertainties that may exist along the length and breadth of the supply chain. There are several players associated with the supply network which is large and it becomes difficult to coordinate with all of them together in order to deliver the goods. A very important point of note is that though there may be cultural, political, geographical as well as historical differences between the collaborators, the need for collaboration as well as specialization of tasks allotted to them needs to be there in place in the humanitarian supply chains (Van Wassenhove 2006).

The most notable differences between the two are that commercial supply chains are based on products, while humanitarian supply chains are based on people, and the commercial supply chains work for profit while the humanitarian supply chains are not for profit.

Table 4.1 gives a description of commercial supply chains vs humanitarian supply chains.

4.5 Effect of Supply Chains on Daily Lives and the Effect of COVID-19 on Supply Chain

Generally, the importance of supply chains and how they touch our daily lives on a continuous basis are unknown to the common man and it is simply taken for granted. Supply chains are rightfully called as the central nervous system of the business

Table 4.1 Difference between commercial and humanitarian supply chain

S. No.	Commercial supply chain	Humanitarian supply chain
1	Business-oriented and designed to make profits	People-oriented and designed to serve people
2	Expected to be present all through the year	Expected to be present only when there is a need (disaster including epidemics and pandemics)
3	Demand is relatively stable and predictable	Demand is unstable and unpredictable
4	Quantity and time of demand can be estimated	Quantity and time of demand cannot be estimated
5	Reaches through middlemen	Generally no middlemen involved. Mostly volunteers and government staff are involved
6	Well-defined and well-planned methods are used for determination of level of inventory where lead time, demand for product and target customer and level of service is the basis	Inventory control is tough and limiting because lead times, demand and locations of demands are highly variable in nature.
7	Middlemen determine the lead time	Material flow determines the lead time
8	Fixed and well-defined network	Unknown network
9	No last mile adjustments done	Many last mile adjustments required
10	Profit maximizing is the goal	Saving lives and reaching people are the goals
11	High quality and low cost is the objective	Reaching as quickly as possible and to as many as possible is the objective
12	Performance is gauged based on resource	Performance gauged based on output
13	Reliable data and information are available	Unreliable data and information or no data and information available
14	End user pays	End user receives without paying
15	Products received from known sources	Products received from unknown sources

economy. All the transport vehicles we see on the roads, the trains on the rail tracks, the flights in the sky and ships in the seas are all the physical manifestations of the supply chains that are making life easier for us and helping us get the things we need at the desired time.

Customers have started having high expectations in terms of the products reaching them on time and in the required quantities at required locations because their works depend on the timely arrival and availability. Further, the supply chains have started playing a vital role with respect to the society. The societal role of the supply chain is so important in today's world that we can say that the very existence of human life today is dependent on supply chains as they deliver the basic necessities such as food and water to all parts of the world. Moreover, supply chains play a major role in the supply and distribution of healthcare products, such as medicines, vaccines, gloves, chemicals, lifesaving healthcare equipment etc., to save a large number of lives on a daily basis. Supply chains have a profound effect on the economy of a country. It can be observed that the developed nations in the world are the nations that have the best infrastructure too, such as excellent roads and road connectivity, vast tracks of railroads, the best airports and excellent seaports. These help goods and supplies to reach consumers at the shortest possible time at the lowest possible costs. On the other hand, it may be observed that the underdeveloped nations are the nations that lack such supply chain infrastructure. When products can reach consumers at low cost in a short span of time, it helps customers to buy more products and thereby increases their standard of living as well due to the availability of more income. It either acts as a lever to economic development or ends up as an impediment to the economic development. Several jobs have also been created in the field of supply chain across the world to operate and manage all the supply chain related activities, such as transporting, warehousing, inventory, packaging delivery etc.

With the onset of corona virus the governments of various countries started taking drastic control measures to curb the spread of the corona virus infection across the globe. The measures included full lockdowns and partial lockdowns which resulted in the closure of several industries temporarily for a certain period of time and later they were allowed to commence production with partial attendance of employees. This led to stoppages of production in many cases and decreased production of products in case of some industries. This led to the unavailability of many products for the customers. To add to the woes of the customers, the retailers were not allowed to be open for business, making it all the more difficult for customers to have access to products. This demand and short supply resulted in increase of prices of products exorbitantly. The Institute of Supply Chain Management of USA conducted a survey and had given the following that as on 25 March 2020,

- More than half of the Tier 1 firms that sources components from China had longer lead times compared to the previous year
- Chinese manufacturers were operating at 50% production capacity, and 56% employees were only involved in the production
- 23% companies reported that they were facing supply chain disruptions
- 62% respondents reported delayed orders from China
- more than half of the companies faced difficulty.

As each and every day passes by, we are able to feel the repercussions of the outbreak of the corona virus being felt. Despite several efforts to curb the infections and the spread of the virus, the onslaught of the virus remains unabated, thus resulting in loss of life and loss of livelihoods as well. China has been considered as the de facto factory of the world and the temporary closure of several factories in China to control the spread of coronavirus had resulted in several disruptions in the supply chain and caused severe damage to the economy, which the world has not faced of late.

4.6 Challenges Faced by Humanitarian Supply Chains

This topic is meant to provide key insights into the various challenges and limitations of the humanitarian supply chains with a special reference to the COVID-19. The government as well as relief organizations involved in humanitarian relief provisions are faced with several unprecedented challenges when disaster strikes. Generally, humanitarian supply chains are circumstances which require very quick responses to the needs that are unforeseeable and this makes it difficult to plan properly and provide support for actual demands (Bag et al. 2020). The most common challenge faced is in terms of mobilization of resources. Generally, men, material, money, machines and information would be mobilized and brought together from various parts of the state/country as per the need of the hour from the places they are available. Further, humanitarian organizations are facing more challenges to improve their efficiencies and effectiveness in their efforts. The reasons for these challenges are due to lack of trust, bad collaboration and incompetent response ability when it is needed (Dubey et al. 2020).

But the corona virus pandemic has created a completely unique and different situation. In normal conditions of distress and disasters such as earthquakes, volcanic eruptions, floods, wildfires or hurricanes, goods produced elsewhere would be rushed to the areas of requirement as and when required as the distress and disaster is confined to a particular local area only and the resources required to mitigate the issues are mobilized from the surrounding areas or from the nearest possible areas. That is, when Chennai in Tamil Nadu, India suffers, the areas around Chennai do not suffer and essentials can be sourced from nearby towns or cities. However, pandemics and epidemics are global in nature. They know no boundaries and borders. Therefore, the efforts to fight against pandemics and epidemics like the present COVID-19 requires a large amount of coordinated teamwork and collaboration of every individual. There is no better time than a pandemic for the entire world to come together and act as one. We need to be reminded that if we do not do so, the result is going to be entirely catastrophic. Almost every country and every state in these countries is faced with several issues related to the COVID-19. Due to lockdowns imposed by the governments of countries and states, the production of several essential items required for humanitarian causes had come to a halt. When there is no availability of essential goods, where is the question of the goods reaching the people in distress? Hence availability of essential commodities became scarce. This led to inability to provide humanitarian assistance. To add fuel to fire, transportation services such as trains, air cargo, sea cargo, public transport etc. had been stopped at several places leading to inability to transport the least available commodities too. Such challenges had added to the already existing woes in the form of price hikes due to the demand making it prohibitively expensive to provide relief assistance. This was not the case with just the developing countries with dense populations but it was the same case even with the developed nations across the globe.

The COVID-19 is not only a humanitarian crisis but also a health crisis too. This is a reflection of a unique crisis that mankind has not witnessed before. Hence the challenges are also expected to be unique.

Given below is a list of challenges faced due to the prevalence of COVID-19.

4.6.1 Demand Uncertainty

The type of demand and the quantity of demand is completely unknown. Since the consumers are unknown, their needs and requirements are also unknown. The demand for the products keeps varying depending on the intensity of the impact of the crisis. How much of which product/service is needed is highly unpredictable. This may vary depending on the population size and the population density in the particular area affected.

4.6.2 Lack of Training

The type of crisis faced by people may not be the same every time. It keeps varying depending on the type of crisis and intensity of the crisis, population affected etc. Hence even if the product/service providers are trained, they may not be effective to handle all kinds of situations and crisis. Further there is no standard training module that can be effective against all crises.

4.6.3 Lack of Skills

Teams deployed for emergency response during crisis are required to make assessments of the situation and collect data and information of requirements of consumers. There is always a need for rapid response which is accompanied with both physical and mental stress for the volunteers. They are faced with unknown and harsh fields, cultures, people, political situations, rough weather etc. These complicate the ability to execute the intended relief work.

4.6.4 Lack of Technology

In today's global world, commercial supply chains are widely supported by technology to respond to the needs of the customers and they are heavily dependent on technology. But in case of humanitarian supply chain, due to the uncertainty and lack of consistencies in the crisis, technology is not able to help the volunteers to a large extent as they have to be customised depending on the requirement and the customization takes time which is scarce during the crisis and becomes obsolete after the crisis to be used again for another crisis.

4.6.5 Lack of Predictability

The needs of the consumers during a crisis is highly unpredictable. The demand for products needs to be assessed only after reaching the spot and analysing the situation. Demand spikes can be felt several times during the relief operations. The needs could be communicated only after the analysis and the requirements can be mobilized.

4.6.6 Lack of Constant Supply

Since the demand cannot be predicted it is very difficult to provide consumers with a continuous supply. Further, the products are not manufactured to cater to orders. They are supplied by donors and the governments based on anticipated needs only. Moreover, if more than what is required is supplied, the products will not be consumed, and will get wasted, especially perishable items and if there is an under supply then the requirements are not fulfilled. Hence it has always been felt that supply cannot be constant and continuous.

4.6.7 Politics and Culture

If the areas are politically and culturally sensitive, there is a high likelihood of humanitarian aids being politicized and rejected due to cultural indifferences. And criticism and dissatisfaction is likely to be high. Consumers and political parties would complain of inefficiencies which are likely to lower the morale of the relief works resulting in lack of willingness to serve such sensitive areas. The human factors have a huge role to be played in any humanitarian aid, which includes politics, language and customs. If criticized the response operations are subject to severe stress during execution of relief operations.

4.6.8 Efficiency and Success Measurement

In commercial supply chains the efficiency and success of the supply chain can be measured by parameters such as time and customer satisfaction along with profits attained. But in case of humanitarian supply chain the efficiency and success lies in the ability to reach the maximum number of products to the maximum number of people. Unfortunately, the number of people who require the product/services are unknown and cannot be calculated accurately too. Hence the efficiency and success cannot be determined.

4.6.9 Lack of Visibility

Visibility along the length of the supply chain called end-to-end visibility is completely absent in the humanitarian supply chain. On one end are the suppliers, viz., donors and government, and the volunteers are at the other end. People at the other end do not know where the products are until it reaches them. Hence the lack of information about the products in transit results in lack of visibility in the supply chain.

4.6.10 Presence of Opportunists and Fraud

A few of the people who play a supporting role for money play havoc by charging exorbitantly for services as they look at the humanitarian crisis as an opportunity to make money. To add fuel to fire some insane opportunists indulge in fraud using trust and inabilities as the platform to loot. These factors result in breach of trust and inefficiencies in the supply chain process.

4.7 Causes for Humanitarian Supply Chain Disruptions

Today's world is highly interconnected and this interconnectivity is associated with several risks. According to Roberta et al. (2015), the suppliers are subjected to more work and stress when the humanitarian supply chain is expanded geographically. The probability of disruption increases when the supply chains are magnified and control over the activities become more difficult. Humanitarian organizations generally have a large employee attrition rate and do not have the required training in a professional way regarding planning of emergency response and inefficient usage of technology (Sandwell 2011) and the supply chain planning is neglected (Taylor and Pettit 2009)). The disaster response performance would be much better if the preparedness for disasters is good (Van Wassenhove 2006).

4.7.1 Lack of Information about Demand

Disruptions in all facets of supply chains and lockdowns in several parts of the world have resulted in non-availability of information of demands for products and closure of several communication channels temporarily. This has resulted in an information blackout about the demands.

4.7.2 Politics of the Local Government

Politicians are notoriously known for the games they play and utilization of opportunity to garner votes and create a good image for themselves and get a good name. They generally try to take credit for all the good deeds and blame the opposition for all the misdeeds. This results in unprecedented delays in products reaching the needy people on time.

4.7.3 Corruption and Fraud

A few dirty minded and highly self-centred individuals who never care for others end up committing frauds and try to make money out of every opportunity they get. They try to fix commissions and seek bribes for things such as permissions for relief materials distribution etc. Corruption and specifically, bribery can result in permissions to supply substandard products to customers causing grave dangers to the health and lives of the customers.

4.7.4 Lack of Product Availability

Humanitarian supply chains are faced with a need for several products at a very short notice and it would be forced to cater to the needs with limited resources and compromised infrastructure. When disaster strikes, the suppliers of products see it as an opportunity to make profits and end up hiking the prices, thereby making the products more dear. Even artificial demands are created to cash in on available short-term opportunities.

4.7.5 Lack of Funds

To keep the engine chugging and alive, one of the most important resources required is the availability of funds. The need and importance of funds is very evident from the request for donations toward the Chief Minister's and Prime Minister's relief fund every time there is a disaster anywhere in the country. Funds are required to mobilize resources and products from suppliers to the needy and the humanitarian volunteers for their basic needs.

4.7.6 Lack of Experience and Expertise

Most of the people involved in the relief operations are volunteers who come together to help the suffering people only when disaster strikes. They are not professionals and hence lack professional experience and the expertise to carry our relief activities as they are supposed to be carried out. The lack of experience and expertise results in inefficient supply of deliverables to the needy people.

4.7.7 Lack of Volunteers and Donors

Volunteers and donors have a major role to play in the fulfilment of a humanitarian supply chain and non-availability of volunteers and donors can cause severe disruptions in the humanitarian supply chains. Though this may not happen many a times, there is a small possibility of non-availability of donors and volunteers. Non-availability of donors will make it difficult to procure products for disbursal, and non-availability of volunteers will affect the distribution and execution process.

4.8 Preparedness of Humanitarian Supply Chain

Disasters have always been sudden and have an augmented occurrence which ends up pressurising the organizations that provide relief activities (Azmat et al. 2019). According to the International Federation of Red Cross (IFRC), the preparedness for mitigating and handling disasters refers to the steps taken to be ready and anticipate the effects of disasters.

Though several attempts have been made in the field of humanitarian supply chain, management and disaster response, we have a long way to go to reach good standards of providing response as we are faced with the limitation of inability to properly identify and structure the challenges in humanitarian supply chains (Sahebi et al. 2017).

It is the prediction ability that plays a vital role whenever there is a possibility of disaster prevention and mitigation of disaster impacts, especially on the vulnerable population, to respond on time whenever required and cope with the impact effectively.

IFRC further states that preparedness for disaster is continuous in nature and that the process is integrated, which is a result of a variety of means for reduction of risks and resources, instead of being a set of distinct activities by itself. The contributions are required from several areas, ranging from logistics, training, healthcare, recovery, livelihood, institutional development etc.

Being prepared becomes a platform for designing and coordinating planning activities as effectively as possible. Preparedness not only helps in effective implementation of the efforts put in for response but also reduces duplication of work and prevents reinventing the wheel. Preparedness cannot emanate from one single source but it is a result of contributions linked to all areas of the humanitarian supply chain. Preparedness can result in saving several lives and the livelihood of many people and help people return to their normal life as quickly as possible in the shortest possible time period.

Generally, the humanitarian supply chain is considered to be unique but the humanitarian supply chain in the COVID-19 context is considered to be even more unique because several restrictions have been imposed by the governments to curtail the spread of the virus. Generally, the HSC operates on projects that are critical missions, where time is considered to be the most important factor. Hence preparedness is an important factor that needs to be taken into consideration for handling the COVID-19 pandemic. During the preparatory phase of a pandemic a contingency plan has to be installed and adopted. The key stakeholders for a humanitarian supply chain are donors, logistics providers and volunteers. When these three important stakeholders are prepared, the response is much better and effective, but unfortunately we also have to accept the fact that crises and pandemics do not announce their arrival but come in so fast and suddenly that preparation cannot be fulfilled. The response phase takes over as after the disaster strikes and preparation for response becomes more important than anything else. Hence it is important that we invest more in the preparation phase to tackle uncertainties and disasters better. Pettit et al. 2011, through their studies have revealed that more lives could be saved by being well prepared during the preparation stage rather than investing materials and money during the operational stages of the efforts related to relief. According to Van Wassenhove (2006), there are claims that 60–80% of the disaster relief work comprises logistics, and the easy way to achieve goals during such efforts is by providing logistics operations and supply chain management systems that are effective and efficient. Hence capacity building of humanitarian logistics and attaining operational effectiveness have become the need to form the needed foundation for humanitarian efforts of relief undertaken by the international community, and according to Thomas (2005), the ability of the logisticians to support buying, transportation and receive supplies at the place where humanitarian relief is required determines the response time of the humanitarian aid after disaster strikes. Preparation can be done with the help of a plan where a plan is a set of activities that need to be executed, including what works need to be carried out, who will be responsible for what activity and how the resources will be mobilized from where. In order to be prepared, plans need to be in place at the local, regional, state and national level. During preparation several factors have to be taken into account, such as expected demand, logistic and infrastructure support, ability and capability to respond etc. Preparedness needs to also include a contingency plan for the already made plan. Other things that need to be considered while preparing for tackling disasters and pandemics is standard operating procedures (SOP) and standard response procedures (SRP). The Governments generally have their own lists, responsibilities, SOPs, SRSs, priorities, action plans etc. and so do humanitarian organizations. Preparedness is all about having a response system in place. However, one needs to understand that 100% preparation is impossible but at the same time, planning and preparation helps in providing better responses in times of need.

4.9 Discussion

With the current COVID-19 pandemic playing havoc across the globe, it has resulted in shortages of supplies of very basic needs to both live and save lives of people. Scarcity of food, medicines and healthcare equipment has started to surface due to the highly infectious nature of the corona virus. A very important point of note here is that the pandemic has led to disruptions of several supply chains, which include both the commercial supply chains and humanitarian supply chains. The main cause of disruptions is completely man-made. In order to curb the spread of the infections, several restrictions had been imposed on the commercial supply chains, which in turn has taken a toll on the humanitarian supply chain as well. This chapter has tried to explain what a humanitarian supply chain is in comparison to the commercial supply chain and how they function. The various challenges faced and the reasons for the disruptions of the humanitarian supply chain have been discussed in this chapter. Discussion related to proper planning and execution of the plans and the limitations and constraints related to the supply of materials by the humanitarian supply have been discussed along with the preparedness levels that need to be in place to handle the problems arising due to the pandemic. Also, the ways and means need to be chalked out to mitigate the shortages during the pandemic times have also been mentioned.

4.10 Conclusion

The humanitarian supply chain is very essential part of the existence of humans, and the management of the humanitarian supply chain is a very complex process subject to constraint by several limitations, such as availability of supplies, availability of volunteers, availability of machines etc. The emphasis needs to be on effective and efficient mobilization of supplies, resources to take the supplies, and the required people at the time of need in required quantities at the required time in the quickest possible way at the least possible cost. In order to execute the humanitarian supply chain process effectively and efficiently, planning and preparation needs to be done perfectly. Since preparation and planning during a pandemic cannot be perfect, it is the need of the hour to keep the stakeholders always ready with the best possible plan and the maximum possible preparedness to save lives as much as possible.

References

Azmat, M., Atif, M., & Kummer, S. (2019). Identification and prioritization of critical success factors in faith-based and non-faith-based organizations' humanitarian supply chain. *Journal of International Humanitarian Action, 4,* 20.

Bag, S., Luthra, S., Venkatesh, V. G., & Yadav, G. (2020). Towards understanding key enablers to green humanitarian supply chain management practices. *Management of Environmental Quality, 31*(5), 1111–1145. https://doi.org/10.1108/MEQ-06-2019-0124.

Balcik, B., Beamon, B., Krejci, C. C., Muramatsu, K. M., & Ramirez, M. (2010). Coordination in humanitarian relief chains: Practices, challenges and opportunities. *International Journal of Production Economics, 126,* 22–34.

Beamon, B., & Balcik, B. (2008). Performance measurement in humanitarian relief chains. *International Journal of Public Sector Management, 21*(1), 4–25.

Dubey, R., Bryde, D. J., Foropon, C., et al. (2020). Agility in humanitarian supply chain: an organizational information processing perspective and relational view. *Annals of Operations Research.*

Ergun, O., et al. (2009). Dagstuhl Seminar Proceedings 092611 Models and Algorithms for Optimization in Logistics. http://drops.dagstuhl.de/opus/volltexte/2009/2181.

Kunz, N., & Gold, S. (2017). Sustainable humanitarian supply chain management—exploring new theory. *International Journal of Logistics Research and Applications, 20*(2), 85–104.

Petrudi, S. H. H., Tavana, M., & Abdi, M. (2020). A comprehensive framework for analyzing challenges in humanitarian supply chain management: A case study of the Iranian Red Crescent Society. *International Journal of Disaster Risk Reduction, 42,*.

Pettit, S., Beresford, A., Whiting, M., & Banomyong, R. (2011). The 2004 Thailand tsunami reviewed: Lesson learned. In M. Christopher & P. Tatham (Eds.), *Humanitarian Logistics: Meeting the challenge of preparing for and responding to disaster* (pp. 103–120). London: Kogan.

Roberta, S., Russel, J., Hiller, S. (2015). Applying best supply chain practices to humanitarian relief. In: Proceedings of the ISCRAM 2015 conference—Kristiansand.

Sahebi, I. G., Arab, A., & Moghadam, M. R. S. (2017). Analyzing the barriers to humanitarian supply chain management: A case study of the Tehran Red Crescent Societies. *International Journal of Disaster Risk Reduction.*

Sandwell, C. (2011). A qualitative study exploring the challenges of humanitarian organisations. *Journal of Humanitarian Logistics and Supply Chain Management, 1*(2), 132–150.

Taylor, D., & Pettit, S. (2009). A consideration of the relevance of lean supply chain concepts for humanitarian aid provision. *International Journal of Service Technology and Management, 12*(4), 430–444.

Thomas, A. (2005). Improving aid effectiveness: Two studies suggest solutions. UN Chronicle 4 (pp. 61–63). http://www.fritzinstitute.org/PDFs.

Van Wassenhove, L. N. (2006). Blackett Memorial Lecture - Humanitarian aid logistics: supply chain management in high gear. *Journal of the Operational Research Society, 57*(5), 475–489.

Chapter 5
Supply Chain Adoption During COVID-19 Lockdown: Global Scenario

Lenin Nagarajan, Siva Kumar Mahalingam, Jayakrishna Kandasamy, and Selvakumar Gurusamy

Abstract This chapter presents an assessment of the implications of the COVID-19 pandemic for supply chains and its flexibility to face the demand of products needed by the general public. The disruption caused by the COVID virus occurrence have created the abrupt impacts on consumer behavior and their preferences. Due to these impacts, consumers are now increasingly exercising care on what, where, and how they buy. The panic buying behavior of the consumers also creating the shocks on demand of products. The effects of demand-side shocks on the supply chains are discussed in this chapter with respect to key items like food and medicines and the rapid change in consumption patterns. Potential supply-side disruptions to supply chains are assessed, including the ingredients to make foods and medicines, workers scarcity, disruptions to transportation, and the issues related with supply of the products between the states and the countries. Further, this chapter addresses long-lasting effects of COVID-19 pandemic on the nature of supply chains, including the growth of the online products delivery sector, and the extent to the supply chain adoption by the priority of "local" food supply chains during COVID-19 lockdown.

Keywords COVID-19 · Panic buying behavior · Demand-side shocks · Supply-side disruptions · Supply chain adoption

L. Nagarajan (✉) · S. K. Mahalingam
Mechanical Engineering, Vel Tech Rangarajan Dr. Sagunthala R&D Institute of Science and Technology, Chennai 600 062, Tamil Nadu, India
e-mail: drleninn@veltech.edu.in

J. Kandasamy
School of Mechanical Engineering, Vellore Institute of Technology, Vellore 632 014, Tamil Nadu, India

S. Gurusamy
Mechanical Engineering, SSN College of Engineering, Chennai 603 110, Tamil Nadu, India

A. R. Sakthivel et al. (eds.), *Managing Supply Chain Risk and Disruptions: Post COVID-19*, Management and Industrial Engineering,
https://doi.org/10.1007/978-3-030-72575-4_5

5.1 Introduction

The COVID-19 pandemic has disclosed the requirement of growth of supply chains to satisfy the fluctuations in demand of food and medicine in order to minimize risk. It looks that this pandemic has created situations in which all movement of products has stopped. Hence, this pandemic can be compared with world war situation (Seric et al. 2020). Though the numerous resemblance between these two, there may be few disparities like residing-in rather than called out to facilitate, and a lockdown on businesses rather than increasing production. However, both pandemic and world war situations changed the business landscape abruptly, especially the businesses on food and medicines. During this pandemic, the supply chains have been required to adjust quickly to handle the demand-side shocks due to the purchasing and consumption pattern, panic buying, in addition to handling the supply-side disruptions due to resource and worker's scarcity and trouble in transportation and product's supply systems (Hobbs 2020). This chapter addresses the challenges that food and medicine supply chains faced globally during the present disaster and provides the strategies and policies to improve the adoption of supply chain. During this chapter writing, the world is presently in the middle of the pandemic. It is more difficult to address all the implications without a clear picture about length and breadth of the pandemic. This chapter addresses an early assessment of few potential implications for food and medicine supply chains.

5.2 Issues on Demand Side

The major issues on demand side have been identified from the consumer-driven businesses by interacting with the retailers and consumers with respect to supply chain perspective. Important issues are detailed below.

5.2.1 Uncertainty from Day-Zero

The real-time requirement of consumers is to be analyzed for the procurement of food items and medicines by retailers under multiple scenarios. Especially during this COVID-19 pandemic, the retailers of consumer goods like food and medicine are facing out-of-stock situations, as consumers are reacting to this uncertainty by excess buying. Day-zero means the date from which the lockdown restrictions implemented on business and on consumer's movement toward purchase (Razdan 2020). It is quite difficult to predict and manage the stocks of food and medicines due to this uncertainty in the purchase by consumers. The supply chain planning and stock levels of those critical items after day-zero is also difficult due to the lockdown restrictions.

5.2.2 Panic Buying Behavior of Consumers

The clear picture in the starting phase of the COVID-19 pandemic has disclosed the unexpected demand created on the food items and medicines rather than other consumable items due to the panic buying behavior of the consumers (Mukandwal 2020). Because those items are vital to lead the day-to-day life by the consumers. If any shortage in these items may lead to increase in the cost of buying and also it cannot fulfill the real needy people. In another aspect, non-practicing of the social distancing between the consumers during shopping due to their panic buying behavior and movement restrictions by the government will lead to the chance of COVID-19 infection among consumers.

5.2.3 Failure in the Policy of Retailers

Usually, most of the retailers and distributors are not interested to invest huge amount of money in the form of stocks even though those items are very essential to the consumers, like food and medicines (Mussell et al. 2020). Nowadays, they are following just-in-time concept for getting delivery from the suppliers. This will not be applicable for pandemic situations, since the quantity of purchase is very small in numbers. Further, the supply chain restriction due to lockdown may even affect the routine supply of these items. Hence, the retailers and distributors may opt the suitable policies to procure these essential items during this kind of pandemic scenario. Accordingly, they should make the suitable contract with the suppliers to face the unexpected demand. In addition, the ration strategies may be adopted by the retailers for the selling of essential products to their regular customers.

5.2.4 Pressure on Retailing Sectors

Globally, it is identified that the major consumption of food is being from the restaurants, hotels, bars, and cafes. Due to the COVID-19 pandemic lockdown restrictions, most of the people are working from home and they are preparing their own foods at home. They are depending on the retailing sectors for the purchase of food ingredients. It will create the unintended pressure on the retailing sectors, so as to satisfy the regular and as well as the new consumers.

5.2.5 *Income of the Consumers*

The economy all around the world is falling down due to the lockdown restrictions on the businesses and the inoperation of private and government sectors. Considerable number of employees of private concerns are laid off from their work and few concerns have cut-off their employee's salary. This sudden fall in consumer's income creates an impact on the demand pattern of the essential food items and medicines. Probably, these type of consumers prefer the low-cost products, instead of buying expensive products with good quality. This phenomenon is directly affecting the procurement of food and medicines by the retailers from the suppliers and the subsequent supply chains also.

5.2.6 *Safe Trading*

Safe trading is one of the immediate requirement by the consumers rather than product's cost. COVID-19 pandemic has created an awareness to the consumers for giving priority to the safety and health aspects rather than the cost and convenience (Brace 2020). In this pandemic situation, the consumers are very keen to buy the products from the particular companies that are taking valid steps to ensure the control in the spread of COVID-19 virus within their span of operations, and further extent. Hence, all the companies must ensure safe trading of products to the consumers by distributing with minimal touchpoints.

5.3 Issues on Supply Side

The possible issues on the supply-side include procurement of raw materials, product and stock management, labor shortages, and disruptions to transportation (Larue 2020). These issues are critically analyzed and presented in this section.

5.3.1 *Criticality in the Raw Material Procurement*

The assessing of availability of raw material for making the products based on the key terms like high-demand products, high-value products, and profitable products by the retailers is very significant. The capacity of the supplier of raw materials and the instability in the price of raw materials are also equally important factors. Further, the industries are reviewing the agreements made with the suppliers regarding permittable variations in the quantity of supply and the cost. In addition, the review of emergency plans is also required to emphasize the potential risks associated with it.

The outcomes of these reviews will be useful to identify the alternate plans for the addressing the criticality in the raw material procurement and the associated cost of supply chains.

5.3.2 *Lack of Support from the Supplier*

Identification of right suppliers is the foremost task of the consumer-oriented organizations based on the key factors, like ability of the supplier in terms of right quantity at right time, quality of the supply materials, cost of the supply materials, etc. This kind of assessment may provide the level of support offered by the suppliers in pandemic situation. If there is a lack in support from the supplier, it is not possible to supply the essential food and medicine items to the consumers on time.

5.3.3 *Labor Shortages*

The movement restrictions of people/labors imposed by the government due to this pandemic situation and the mentality of labor about self-isolation to prevent them from COVID-19 infection are the key factors of labor shortages. The production rate of all the food processing industries and pharmaceutical industries are mainly dependent on the work of labors. Further, the distribution of these items through supply chains is also dependent on the size of the labors who are really working on it. Hence, this labor shortage created a strong impact on supply side to the retailers/distributors.

5.3.4 *Product's Management*

Few retailers may have the capability to arrange their own supply chains even during the lockdown restrictions, by moving the customized products mix from the supplier's place. In this circumstance, the supplier should identify the sharing of raw materials to manufacture those products mix, facilities to manufacture, etc. The retailers should make an emergency plans along with their suppliers in view of managing the product portfolio to satisfy the need of the consumers on time.

5.3.5 *Stock Management*

Managing the stock levels, reorder levels, and economic order quantity of essential food and medicine items are somewhat difficult to the retailers due to the drastic variation in the purchasing pattern of the consumers (Razdan 2020). This variation in the purchasing pattern is because of the consumer's panic buying behavior with respect to their intuition on the stock-outs. In the other aspect, the support from the supplier side is also equally important in the stock management. Retailers must know the ability of the suppliers and their production schedules, supply chains, etc.

5.3.6 Restriction on Supply Chain

The lockdown restriction on the supply chains between the districts, states, and countries imposed by the authorities is the crucial factor for the timely supply of food and medicines to the consumers. Retailers should assess the feasible locations of delivery centers and distributors to compute risks by geography. Analysis on warehouse locations and their capacity are the key factors in fixing the supply locations. The proper export plans and schedules and the associated risks with the destinations are essential parameters to ensure the better flow of products under constrained supply chain. The partnership business is also one of the good remedy in these circumstances (Cao and Zhang 2011).

5.3.7 Financial Position of Consumer Companies

A careful examination of their financial position by the consumer companies is required to decide the buying capability, stock management, cash flow, etc. They have to meet out the excess expenses toward the variation in the cost of procurement of materials due to sudden rise in demand, restricted supply chains, etc. Further, the inability of the consumers based on their buying capability because of the drastic changes in the price of the products play another important role which affects the cash flow of the consumer companies. Hence, the multiple-scenario analysis by the consumer companies are required to know their financial position to meet out the pandemic situations and further action for financial alternatives.

5.4 Supply Chain Adoption

Most of the manufacturing industries are facing two kinds of issues during this COVID-19 pandemic. The first one is that the lack in the quantity of production due to labor shortages and the second thing is that the supply chain disruptions due to the lockdown restrictions by the government authorities. As a remedial action for the first issue, many of the manufacturing industries are presently looking to robotics technology/automation to supplement locked down labors, reinforce health and security measures, and rescue their production operations.

In view of avoiding the issues in supply chains of food and medicines, globally, the manufacturing industries start preferring the local suppliers for raw material procurement rather than the usual contract suppliers. In the similar way, marketing of the manufactured products is taken care by the local retailers and distributors. Since the government has not imposed any stringent restriction on the supply of essential items like food and medicines, the local supply chain is a viable solution for marketing the products. Further, there is no government restriction to get the essential products from the manufacturing industry by the retailer/distributors. Simultaneously, the government of different countries has allowed few distributors to do the online sale of food and medicine products by ensuring the contactless delivery to the consumers, especially in the urban areas. This exemption by the government is very ease for the consumers to procure the essential items without any panic of COVID-19 infection.

In addition, this is the time for the farmers to sell their products directly to the consumers within the local region rather than selling to the distributors. Approximately 30–40% of the consumer needs are fulfilled through this method of local supply. From this method of supply, the farmers are also benefitted by gaining the direct profit without any distributing commission. These viable methods ensure the self-adoption of supply chain during this COVID-19 lockdown. This self-adoption of supply chain provides the opportunity for the consumers to procure the essential items, and it will contribute to a major percentage to sustain the economy of the world during this lockdown period.

5.5 Conclusion

The world is facing numerous disruptions in the supply chains of essential food and medicine items during this COVID-19 pandemic. The usual just-in-time concept for getting supply on the right time by the manufacturers/retailers/distributors/consumers is not a suitable solution during this pandemic time, since the governments all around the world imposed lockdown restriction. The flexible procurement strategies rather than the fixed strategies are the suitable solution for all. The issues on supply chains related with demand and supply-sides are elaborately discussed as a main content in this chapter. Further, the effective ways to overcome these issues by the supply chains are discussed under the title of supply chain adoption. The local food supply chain, online products delivery sector, and direct supply of agricultural/food products by the farmers are the good instances for supply chain adoption. This self-adoption of supply chain ensures the supply of essential products to the consumers without any shortfall and also helps sustain the economy of the world.

References

Brace, R. (2020). Supporting suppliers through the Covid-19 pandemic. *EuroFinance.* https://www.eurofinance.com/news/supporting-suppliers-through-the-covid-19-pandemic/. Accessed 30 March 2020.

Cao, M., & Zhang, Q. (2011). Supply chain collaboration: Impact on collaborative advantage and firm performance. *Journal of operations Management, 29*(3), 163–180.

Hobbs, J. E. (2020). Food supply chains during the COVID-19 pandemic. *Canadian Journal of Agricultural Economics/Revue Canadienne d'Agroeconomie.*

Larue, B. (2020). Labour issues and COVID-19. *Canadian Journal of Agricultural Economics/Revue Canadienne d'Agroeconomie.*

Mukandwal, P. (2020). Supply chain and pandemics. *USA.* https://www.mtu.edu/unscripted/stories/2020/april/supply-chains-and-pandemics.html. Accessed 22 April 2020.

Mussell, A., Bilyea, T., & Hedley, D. (2020). Agri-food supply chains and Covid-19: Balancing resilience and vulnerability. *Agri-Food Economic Systems.*

Razdan, H. (2020). Ramping up the supply chain post Covid-19. *KPMG in India.* https://home.kpmg/in/en/home/insights/2020/05/ramping-up-the-supply-chain-post-covid-19.html. Accessed 27 May 2020.

Seric, A., Gorg, H., Mosle, S., & Windisch, M. (2020). Managing COVID-19: How the pandemic disrupts global value chains. *UNIDO.* https://iap.unido.org/articles/managing-covid-19-how-pandemic-disrupts-global-value-chains. Accessed 28 April 2020.

Chapter 6
Design and Development of Decision Support System for Emergency Need Relief Distribution

V. Kamala and K. Lakshmi

Abstract This study focuses on the design and development of a Decision Support System (DSS) using DEMATEL and TOPSIS which helps decision-maker in the humanitarian relief chain, to release funds to a disaster victim by considering their economic status. The knowledge base of the proposed DSS consists of a range of economic indicators such as Per capita income of the family, stock available in INR, a number of dependent family members and Asset owned define the ranking of community resident's economic status. DEMATEL is used to find the weightage of Emergence need assessment parameter such as Per capita income of the family, stock available in INR, number of dependent family members and Asset owned. TOPSIS is used to rank disaster victims. The Proposed DSS helps decision-maker involved in relief services, to provide transparent decision effectively. The real world data is used to evaluate the efficiency of the proposed approach. This procedure is based on expert knowledge, rules of thumb and population behaviour. Therefore, this study develops a practical decision support system considering disaster victims financial criteria and MCDM is used to rank the victim which help with the relief distribution in the humanitarian supply chain.

Keywords Relief distribution. FIS, DSS · Economic indicators · Decision making

6.1 Introduction

Disasters may be sudden or slow-onset. A disaster is an event that occurs instantaneously and the effects of disasters have intensified economically, socially and environmentally for the past few decades (Kaur and Singh 2019). Disruptions caused by disaster may be short or long-term disruptions and increasing propagation in the economic conditions of disaster victims (Ivanov 2020). When compared to the last few years the occurrence of natural disasters along with the disaster victims has significantly increased by The International Disaster Database (2012). In order to

V. Kamala (✉) · K. Lakshmi
College of Engineering, Department of Industrial Engineering, Anna University, Chennai 600 025, Tamil Nadu, India

A. R. Sakthivel et al. (eds.), *Managing Supply Chain Risk and Disruptions: Post COVID-19*, Management and Industrial Engineering,
https://doi.org/10.1007/978-3-030-72575-4_6

reduce the effects of the disaster, the fulfillment of disaster management before and after the occurrence of a disaster is required. (Galindo and Batta 2013).

As compared with pre-disaster functions, the need for social service professionals immediately after a disaster is significantly increased. The two important emanate after a disaster is confirming the continuity of services and addressing unmet needs (HHS 2014).

COVID-19 is a pandemic that affected more than 3 million people, and about 9,50,449 individuals lost their lives up to August 2020. Every day the rate of infected people and death rates change. the COVID-19 pandemic is affecting many countries across the world and not only affecting the health crisis alone but also bump millions of families back into poverty. have had a There is a significant impact on global supply chains due to COVID-19 Quarantine condition. It is important to provide disaster relief services in compliance with humanitarian relief chain principles to disaster victims (Javaid et al. 2020). The proper flows of relief items must be provided after a disaster in the humanitarian relief chain. We propose a Decision support system for post-disaster phases.

The relief distribution to the victims is one of the challenging problem in the humanitarian supply chain (Sheu 2007). The failure in relief distribution planning will results in failure in the continuity of services and addressing unmet needs people. If disaster victims have a shortage of survival needs, leads to a calamity. The integration of Information, human resource management and technology utilisation of a few approaches and results will help to decrease the affected people. Therefore, an effective way to manage the humanitarian relief chain is to rank the disaster victims who are in survival needs according to the emergency need assessment parameter. The main purpose of this study is to design a decision support system in order to rank the disaster victims for the management of relief distribution in the humanitarian relief supply chain. Figure 6.1 shows the structure under the study of emergency need assessment of the humanitarian relief supply chain.

6.2 Background

In this section, the studies undertaken in the distribution framework in Humanitarian supply chain management and application of DSS have been discussed.

In recent decades Humanitarian relief distribution has received increased attention. Sheu et al. (2007) proposed a three-layer emergency logistics co-distribution conceptual framework that involves disaster-affected area grouping and relief co-distribution. The application of framework is conducted in earthquake disasters in Taiwan. To evaluate the emergency preparedness and responsibility performances of non-governmental humanitarian relief organisations (NGOs), a hybrid approach, that consists of interval type-2 fuzzy sets, AHP and PROMETHEE are proposed by Celik and Gumus (2016). The study is conducted in Turkey to determine the critical success factors (CSFs) of humanitarian relief logistics management operations.

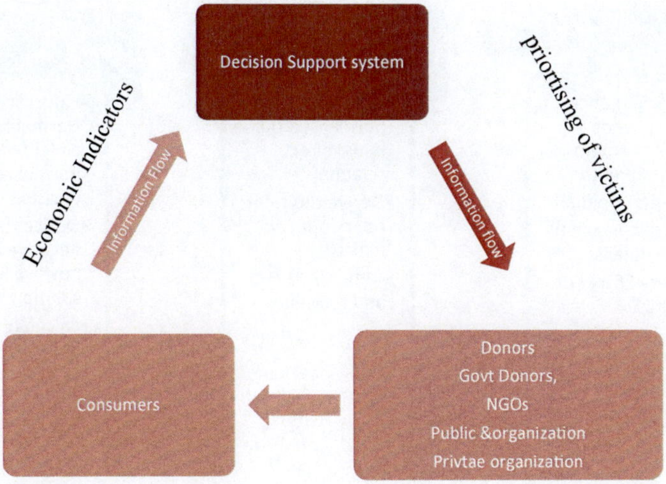

Fig. 6.1 Humanitarian supply chain under study

AHP is used to determine the weightage CSF and PROMETHEE is used to evaluate the preparedness and response abilities of NGOs. Li et al. (2018) developed a maximal cooperative covering model by incorporating the severity of disasters that affect the region, financial efficiency. Maghfiro and Hanaoka (2020) presented a three-level relief distribution model and the location of logistics operational areas, modes of transport utilised, and amount of relief goods allocated for each mode of transport are determined in this model. The factors considered in this study are network, accessibility of supplies, infrastructure conditions and modes of transport. The application of the model is examined using the Yogyakarta Earthquake of 2006 in Indonesia. Similarly, many studies focused on designing the distribution network in the emergency situation (Li et al. 2018; Sahebjamnia et al. 2017).

Govindan et al. (2020) developed a decision support system based on a Fuzzy inference system (FIS) which is used to classify members with the goal to alleviate the pandemic outbreaks for healthcare supply chain disruption. By considering the indicators of age and) The individuals are classified based on age, pre-existent diseases such as high blood pressure, heart problems and diabetes. The proposed approach was verified using the information from four users. Similarly, a decision support system (DSS) is used in different humanitarian supply chain decision such as managing humanitarian relief chains (Sahebjamnia et al. (2017), humanitarian facility location (Florez et al. 2015), logistics of Humanitarian Aid (Ortuño et al. 2010), evacuation location assessment (Hadiguna et al. (2014), disaster relief supply chain (Kumar and Havey 2013).

However, DSS has been developed for a different type of tactical and operational decision involved in the distribution decision of humanitarian relief chains, in this study DSS is developed for emergency need assessment by considering economic

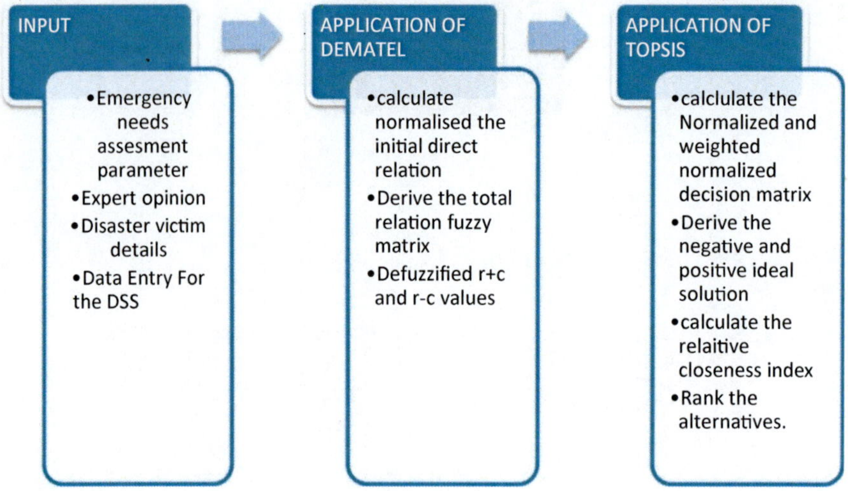

Fig. 6.2 Emergency need assessment—DSS

and welfare parameters to rank the disaster victims for the management of relief distribution.

6.3 Proposed Approach

A decision support system based on the DEMATEL & TOPSIS model is developed to prioritise disaster victims in terms of the economic condition which is shown in Fig. 6.2. The objective of the developed decision support system at assessing the Emergency need parameters in relief distribution of the humanitarian supply chain. The Emergency need parameters considered in this study Per capita income of the family, stock available in INR, number of dependent family members and Asset owned. DEMATEL is used to find the weightage of parameter and TOPSIS is used to rank the disaster victims for the management of relief distribution. Figure 6.2 shows the emergency need assessment—DSS.

6.4 Emergency Needs Assessment Using DEMATEL

An emergency need of disaster victims depends on several factors that are either directly or indirectly related. Hence, assessment of emergency needs based on a single factor alone may not be justifiable. The DEMATEL method can be used to analyse the relationship among multiple complex factors (Tzeng et al. 2007). However, the evaluation criteria in this context may be imperfect or can contain uncertain factors and so the human judgement about the preferences can be difficult to estimate by exact numerical values.

The fuzzy DEMATEL method can help to overcome this and has been used by several researchers. We use the fuzzy DEMATEL method to estimate the criteria weights. The relationship between the factors considered is evaluated based on the opinions of experts using a fuzzy scale. To demonstrate the application of DEMATEL TOPSIS in DSS Numerical calculation is shown below. The fuzzy scale mapping the degree of influence with the linguistic terms is shown in Table 6.1.

The following steps are involved in the fuzzy DEMATEL approach (Visalakshmi et al. 2015).

Step 1: Obtain a $n \times n$ fuzzy direct-relation matrix $\tilde{A}$ from the experts, based on pairwise comparisons of the criteria is shown in Table 6.2. Its elements $\tilde{a}_{ij} = \left(l_{ij}, m_{ij}, u_{ij}\right)$ represent the degree to which criterion j is affected by criterion i.

Step 2: Determine the normalised fuzzy direct-relation matrix $\tilde{X}$ using Eq. (6.1) is shown in Table 6.3.

$$\tilde{X} = s \times \tilde{A} \tag{6.1}$$

Table 6.1 Linguistic terms for fuzzy DEMATEL evaluation

Abbreviation	Fuzzy scales	Linguistic term
N	(0,0,1)	None
VL	(0,0.1,0.2)	Very low
L	(0.1,0.2,0.3)	Low
FL	(0.2,0.3,0.4)	Fairly low
ML	(0.3,0.4,0.5)	More or less low
M	(0.4,0.5,0.6)	Medium
MG	(0.5,0.6,0.7)	More or less good
FG	(0.6,0.7,0.8)	Fairly good
G	(0.7,0.8,0.9)	Good
VG	(0.8,0.9,1)	Very good
E	(0.9,1,1)	Excellent

Table 6.2 Fuzzy direct-relation matrix $\tilde{A}$

	P1	P2	P3	P5
P1	(0,0,0)	(0.6,0.7.0.8)	(0.1,0.2,0.3)	(0.4,0.5,0.6)
P2	(0.6,0.7,0.8)	(0,0,0)	(0.5,0.6,0.7)	(0.2,0.3,0.4)
P3	(0.1,0.2,0.3)	(0.5,0.6,0.7)	(0,0,0)	(0.7,0.8,0.9)
P4	(0.4,0.5,0.6)	(0.2,0.3,0.4)	(0.7,0.8,0.9)	(0,0,0)

Table 6.3 Fuzzy direct-relation matrix (Normalised) $\tilde{X}$

	P1	P2	P3	P5
P1	(0,0,0)	(0.315,0.3684,0.4211)	(0.0526.0.1053.0.1579)	(0.2105.0.2632.0.3158)
P2	(0.315,0.3684,0.4211)	(0,0,0)	(0.2632,0.3158,0.3684)	(0.1053,0.1579,0.2105)
P3	(0.0526,0.1053,0.1579)	(0.2632,0.3158,0.3684)	(0,0,0)	(0.3684,0.4211,0.4737)
P4	(0.2105,0.2632.0.3158)	(0.1053,0.1579,0.2105)	(0.3684,0.4211,0.4737)	(0,0,0)

$$s = \frac{1}{\max_{1 \le i \le n} \sum_{j=1}^{n} u_{ij}}$$

Step 3: Define three crisp matrices based on $\tilde{X}$, where $\tilde{x}_{ij} = (l_{ij}, m_{ij}, u_{ij})$.

$$X_l = \begin{bmatrix} 0 & l_{12} & \cdots & l_{1n} \\ l_{21} & 0 & \cdots & l_{2n} \\ \vdots & & & \vdots \\ l_{n1} & l_{n2} & \cdots & 0 \end{bmatrix}, \quad X_m = \begin{bmatrix} 0 & m_{12} & \cdots & m_{1n} \\ m_{21} & 0 & \cdots & m_{2n} \\ \vdots & & & \vdots \\ m_{n1} & m_{n2} & \cdots & 0 \end{bmatrix}, \quad X_u = \begin{bmatrix} 0 & u_{12} & \cdots & u_{1n} \\ u_{21} & 0 & \cdots & u_{2n} \\ \vdots & & & \vdots \\ u_{n1} & u_{n2} & \cdots & 0 \end{bmatrix}$$

Step 4: Derive the fuzzy total-relation matrix $\tilde{T}$ using Eqs. (6.2) to (6.5)

$$\tilde{T} = \tilde{X}\left(I - \tilde{X}\right)^{-1} \tag{6.2}$$

$$\text{Matrix } \left[l'_{ij}\right] = X_l(I - X_l)^{-1} \tag{6.3}$$

$$\text{Matrix } \left[m'_{ij}\right] = X_m(I - X_m)^{-1} \tag{6.4}$$

$$\text{Matrix } \left[u'_{ij}\right] = X_u(I - X_u)^{-1} \tag{6.5}$$

$$\tilde{T} = \begin{bmatrix} \tilde{t}_{11} & \tilde{t}_{12} & \cdots & \tilde{t}_{1n} \\ \tilde{t}_{21} & \tilde{t}_{22} & \cdots & \tilde{t}_{2n} \\ \vdots & & & \vdots \\ \tilde{t}_{n1} & \tilde{t}_{n2} & \cdots & \tilde{t}_{nn} \end{bmatrix}$$ where, $\tilde{t}_{ij} = \left(l'_{ij}, m'_{ij}, u'_{ij}\right)$ and I is the identity matrix.

Step 5: Defuzzify $\tilde{T}$ using Eq. (6.6) and Table 6.4 show the total influence matrix for each set of sub-criteria considered.

$$F\left(\tilde{t}_{ij}\right) = \frac{1}{2}\int_{0}^{1}\left(\inf_{x \in \Re} \tilde{t}_{ij}^{\alpha} + \sup_{x \in \Re} \tilde{t}_{ij}^{\alpha}\right) d\alpha \tag{6.6}$$

Table 6.4 Total influence matrix

	P1	P2	P3	P5
P1	4.404	5.036	4.951	4.99
P2	5.036	5.183	5.483	5.336
P3	4.951	5.483	5.357	5.63
P4	4.99	5.336	5.63	5.3

Table 6.5 Emergency need parameter weightage

Parameter	P1	P2	P3	P4
Weightages	0.2333	0.2532	0.2577	0.2558

Step 6: Compute the row sum (r) and the column sum (c) of the total influence matrix. $r + c$ gives the degree to which factor i affects or is affected by j. A positive value of $r - c$ means that factor i affects other factors and vice versa if it is negative (Chen and Chen 2010).

Step 7: Determine the weights by normalising the $r + c$ values is shown in Table 6.5.

6.5 Application of TOPSIS

The Technique for Order of Preference by Similarity to Ideal Solution (TOPSIS) is based on the assumption that any chosen alternative should be closer to the positive ideal solution and far away from the negative ideal solution. TOPSIS has the following steps (Chen and Chen 2010).

Step 1: Obtain the decision matrix D for evaluation with elements f_{ij}, based on the various criteria considered is shown in Table 6.6.

Step 2: Determine the normalised decision matrix (Table 6.7) with elements r_{ij} using Eq. (6.7).

Table 6.6 Excerpt of Disaster victim details- Decision matrix

S. No.	Monthly income	Number of dependent family members	Asset owned (INR)	Stock available(INR)
1	4552	2	476256	570
2	9618	5	207199	484
3	9000	5	192583	565

Table 6.7 Excerpt normalised decision matrix

S. No.	Monthly income	Number of dependent family members	Asset owned (INR)	Stock available (INR)
1	0.33	0.27	0.86	0.61
2	0.69	0.68	0.37	0.52
3	0.65	0.68	0.35	0.6

Table 6.8 Weighted normalised decision matrix

S. No.	Monthly income	Number of dependent family members	Asset owned (INR)	Stock available (INR)
1	0.073755	0.068877	0.225664	0.157868
2	0.154215	0.173468	0.097088	0.134576
3	0.145275	0.173468	0.09184	0.15528

$$r_{ij} = \frac{f_{ij}}{\sqrt{\sum_j f_{ij}^2}} \tag{6.7}$$

Step 3: Calculate the weighted normalised decision matrix which is shown in Table 6.8 with elements v_{ij} using the weights w_i for each criteria using Eq. (6.8)

$$v_{ij} = w_i \cdot r_{ij} \tag{6.8}$$

Step 4: Find the ideal and negative-ideal solutions using Eqs. (6.9) and (6.10), where I' is the set of benefit criteria and I'' is the set of cost criteria.

$$A^+ = \{v_1^+, \ldots, v_n^+\} = \left\{\left(\max_j v_{ij} \middle| i \in I'\right), \left(\min_j v_{ij} \middle| i \in I''\right)\right\} \tag{6.9}$$

$$A^- = \{v_1^-, \ldots, v_n^-\} = \left\{\left(\min_j v_{ij} \middle| i \in I'\right), \left(\max_j v_{ij} \middle| i \in I''\right)\right\} \tag{6.10}$$

Step 5: Determine the distance from the positive ideal solution D_j^+ and the distance from the negative-ideal solution D_j^- using Eqs. (6.11) and (6.12) for each alternative j is shown in Table 6.9.

$$D_j^+ = \sqrt{\sum_i \left(v_{ij} - v_i^+\right)^2} \tag{6.11}$$

$$D_j^- = \sqrt{\sum_i \left(v_{ij} - v_i^-\right)^2} \tag{6.12}$$

Table 6.9 Ideal solution

S. No.	D^+	D^-
1	0.131959	0.068877
2	0.130669	0.173468
3	0.134147	0.173468

Table 6.10 Relative closeness index

S. No.	Relative closeness
1	0.507239
2	0.502653
3	0.489028

Step 6: Evaluate the relative closeness to the ideal solution for each alternative *j* and rank the alternatives is shown in Table 6.10 using Eq. (6.13)

$$C_j^* = \frac{D_j^-}{\left(D_j^+ + D_j^-\right)} \tag{6.13}$$

6.6 Design and Development of DSS−Relief Distribution

DSS−Relief Distribution is a web application that was developed using ASP.Net as a presentation layer and Microsoft SQL Server as a database layer. DSS−Relief Distribution consist of interface, knowledge base, model and database. The database consists of decision-makers ratings/opinions and disaster victims' information. The knowledge base consists of conceptual characteristics for ranking disaster victims. The model base includes DEMATEL and TOPSIS approaches used for assessment. DSS−Relief Distribution performs data organisation, decision structure, compare and analysis of TOPSIS output and finding the ranks. The outputs of DSS−Relief Distribution will be the ranking of disaster victims. Figure 6.3 depicts adding,

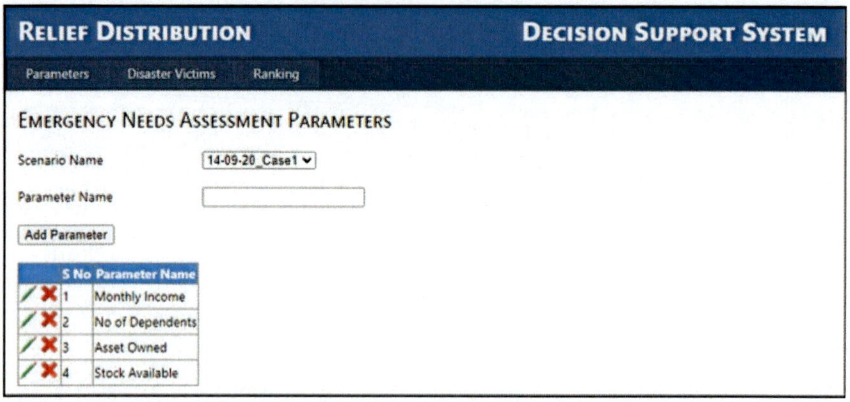

Fig. 6.3 Details of emergency need assessment parameter

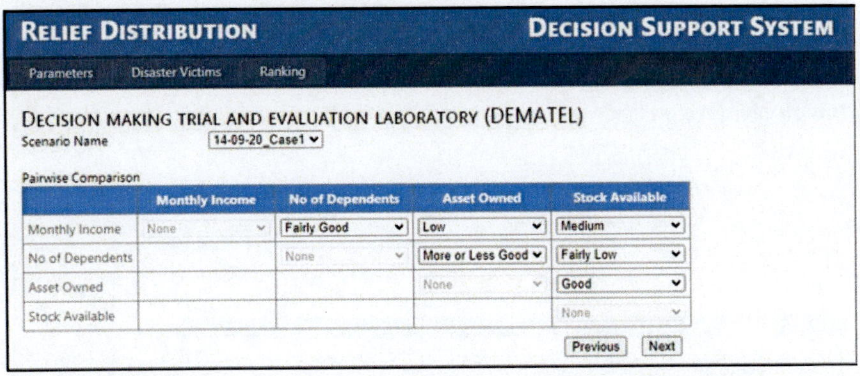

Fig. 6.4 Expert rating

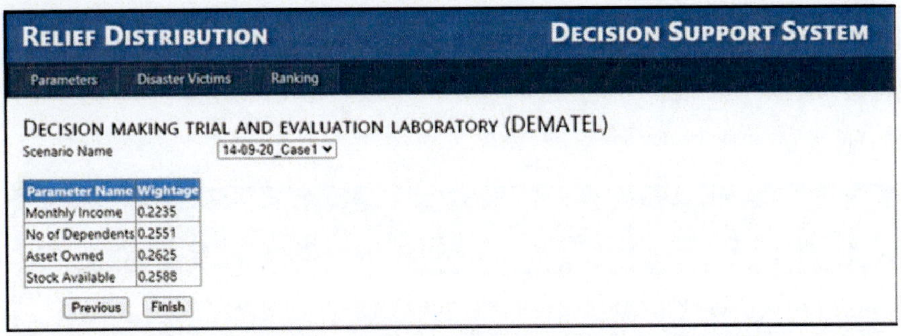

Fig. 6.5 Weightage of emergency needs assessment parameters

editing, deleting and viewing features of the details of emergency needs assessment parameters.

Figure 6.4 depicts the features of updating experts rating based on the linguistic term of the fuzzy scale shown in Table 6.1. The result page of DEMATEL which depicts the weightage of emergency needs assessment parameters is shown in Fig. 6.5.

Adding, editing, and deleting details of disaster victims' features are shown in Fig. 6.6. Figure 6.7 depicts the features selecting parameters and disaster victims' for the TOPSIS model.

6.7 Results and Discussion

Based on DEMATEL and TOPSIS approaches, the relative closeness of each disaster victims' is found. Decision-makers will select the disaster victims' with high relative closeness value. Figure 6.8 depicts the victims identity number (Aadhaar number) and their relative closeness and their rank. Based on relative closeness index calculation

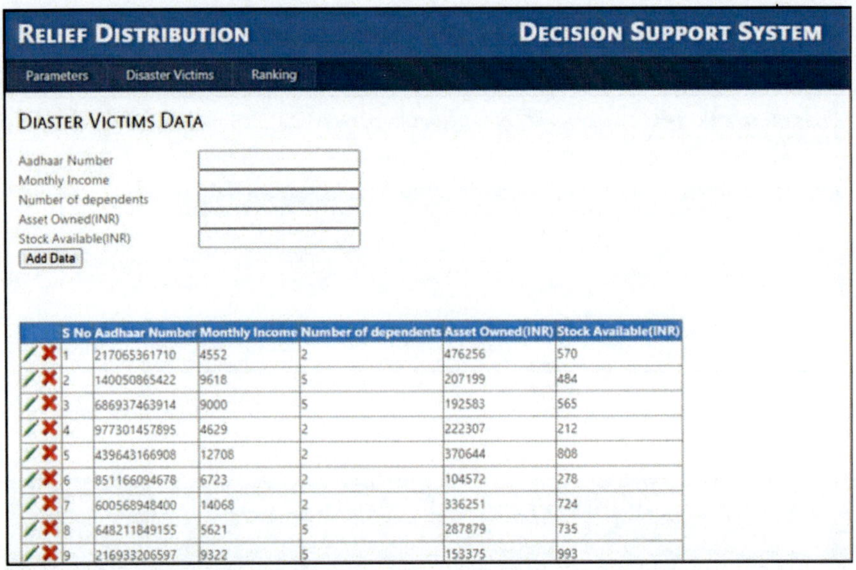

RELIEF DISTRIBUTION DECISION SUPPORT SYSTEM

Parameters Disaster Victims Ranking

DIASTER VICTIMS DATA

Aadhaar Number
Monthly Income
Number of dependents
Asset Owned(INR)
Stock Available(INR)
Add Data

S No	Aadhaar Number	Monthly Income	Number of dependents	Asset Owned(INR)	Stock Available(INR)
1	217065361710	4552	2	476256	570
2	140050865422	9618	5	207199	484
3	686937463914	9000	5	192583	565
4	977301457895	4629	2	222307	212
5	439643166908	12708	2	370644	808
6	851166094678	6723	2	104572	278
7	600568948400	14068	2	336251	724
8	648211849155	5621	5	287879	735
9	216933206597	9322	5	153375	993

Fig. 6.6 Disaster victim details

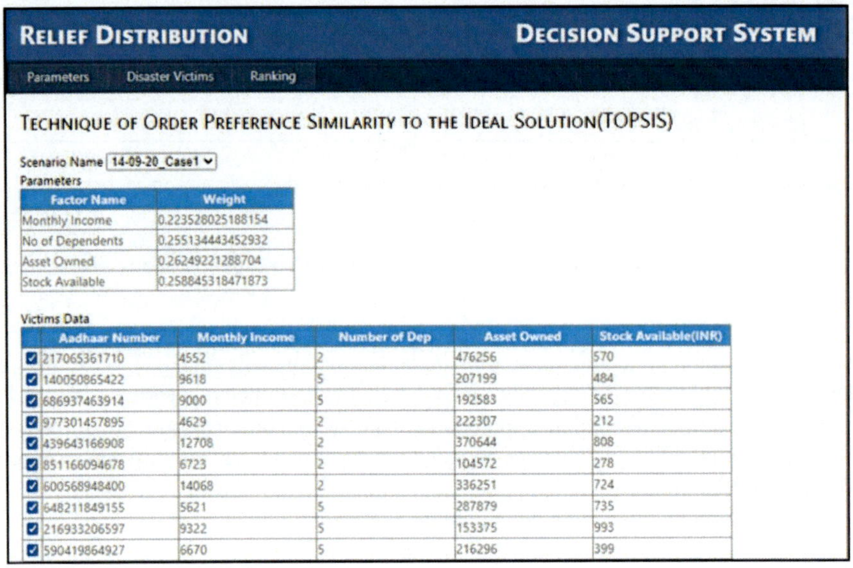

RELIEF DISTRIBUTION DECISION SUPPORT SYSTEM

Parameters Disaster Victims Ranking

TECHNIQUE OF ORDER PREFERENCE SIMILARITY TO THE IDEAL SOLUTION(TOPSIS)

Scenario Name 14-09-20_Case1

Parameters

Factor Name	Weight
Monthly Income	0.223528025188154
No of Dependents	0.255134443452932
Asset Owned	0.26249221288704
Stock Available	0.258845318471873

Victims Data

Aadhaar Number	Monthly Income	Number of Dep	Asset Owned	Stock Available(INR)
217065361710	4552	2	476256	570
140050865422	9618	5	207199	484
686937463914	9000	5	192583	565
977301457895	4629	2	222307	212
439643166908	12708	2	370644	808
851166094678	6723	2	104572	278
600568948400	14068	2	336251	724
648211849155	5621	5	287879	735
216933206597	9322	5	153375	993
590419864927	6670	5	216296	399

Fig. 6.7 Application of TOPSIS

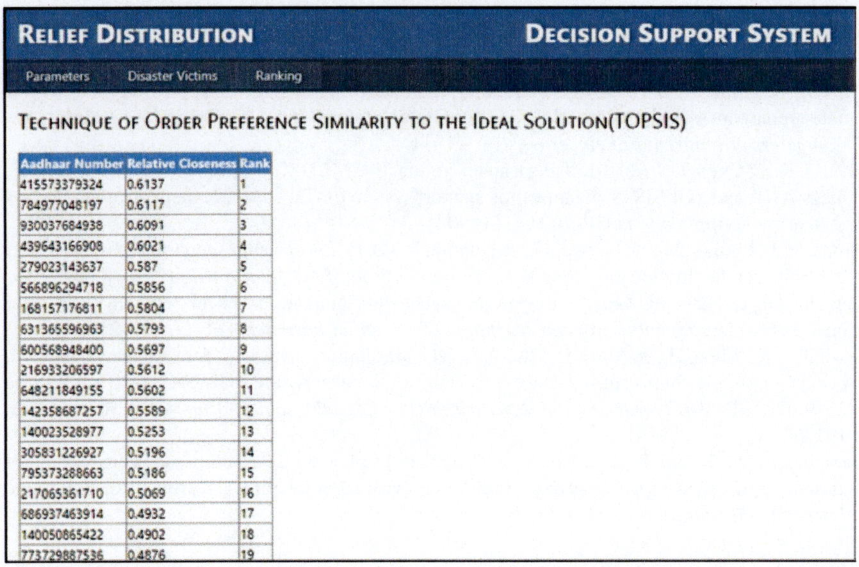

Aadhaar Number	Relative Closeness	Rank
415573379324	0.6137	1
784977048197	0.6117	2
930037684938	0.6091	3
439643166908	0.6021	4
279023143637	0.587	5
566896294718	0.5856	6
168157176811	0.5804	7
631365596963	0.5793	8
600568948400	0.5697	9
216933206597	0.5612	10
648211849155	0.5602	11
142358687257	0.5589	12
140023528977	0.5253	13
305831226927	0.5196	14
795373288663	0.5186	15
217065361710	0.5069	16
686937463914	0.4932	17
140050865422	0.4902	18
773729887536	0.4876	19

Fig. 6.8 Ranking of disaster victims

the emergency need assessment parameter of disaster victim is derived. The ranking of the victims is calculated using relative closeness index values which are shown in Fig. 6.8. The relief distribution DSS helps relief services decision-makers to provide transparent decisions effectively.

6.8 Conclusion

A decision support system was designed to prioritise the disaster victims and, accordingly, manage the relief distribution in the humanitarian supply chain. The disaster victim is prioritised using emergency need assessment parameters such as per capita income of the family, stock available in INR, number of dependent family members, and asset owned. DEMATEL is used to find the weightage of the emergency need assessment parameter and disaster victims are prioritised using TOPSIS. The developed DSS was verified using 30 victims data. Since the computational complexity is more, a computer-generated decision support system has been designed. DSS generates the ranking of disaster victims. The developed DSS enabled the decision-maker to make decisions from the view of relief distribution. the DSS can be developed by adding more evaluation parameters and also evaluated using advanced techniques in the future. The adapted DSS can be tested in many regions to derive meaningful insights.

References

Celik, E., & Gumus, A. T. (2016). An outranking approach based on interval type-2 fuzzy sets to evaluate preparedness and response ability of non-governmental humanitarian relief organizations. *Computers & Industrial Engineering, 101,* 21–34.

Chen, J. K., & Chen, I. S. (2010). Using a novel conjunctive MCDM approach based on DEMATEL, fuzzy ANP, and TOPSIS as an innovation support system for Taiwanese higher education. *Expert Systems with Applications, 37*(3), 1981–1990.

Florez, V. J., Lauras, M., Okongwu, U., & Dupont, L. (2015). A decision support system for robust humanitarian facility location. *Engineering Applications of Artificial Intelligence, 46,* 326–335.

Galindo, G., & Batta, R. (2013). Review of recent developments in or/ms research in disaster operations management. *European Journal of Operational Research, 230*(2), 201–211.

Govindan, K., Mina, H., & Alavi, B. (2020). A decision support system for demand management in healthcare supply chains considering the epidemic outbreaks: A case study of coronavirus disease 2019 (COVID-19). *Transportation Research Part E: Logistics and Transportation Review, 138,* 101967.

Hadiguna, R. A., Kamil, I., & Delati, A., et al. (2014). Implementing a web-based decision support system for disaster logistics: A case study of an evacuation location assessment for Indonesia. *International Journal of Disaster Risk Reduction, 9,* 38–47.

HHS. (2014). *Disaster Human Services Concept of Operations (CONOPS).* Washington, DC.

Ivanov, D. (2020). Predicting the impacts of epidemic outbreaks on global supply chains: A simulation-based analysis on the coronavirus outbreak (COVID-19/SARS-CoV-2) case. *Transportation Research Part E: Logistics and Transportation Review, 136,* 101922.

Javaid, M., Haleem, A., Vaishya, R., et al. (2020). Industry 4.0 technologies and their applications in fighting COVID-19 pandemic. *Diabetes & Metabolic Syndrome: Clinical Research & Reviews, 14*(4), 419–422.

Kaur, H., & Singh, S. S. P. (2019). Sustainable procurement and logistics for disaster resilient supply chain. *Annals of Operations Research, 283,* 309–354. https://doi.org/10.1007/s10479-016-2374-2.

Kumar, S., & Havey, T. (2013). Before and after disaster strikes: A relief supply chain decision support framework. *International Journal of Production Economics, 145*(2), 613–629.

Li, S., Ramshani, M., & Huang, Y. (2018). Cooperative maximal covering models for humanitarian relief chain management. *Computers & Industrial Engineering, 119,* 301–308.

Maghfiro, M. F. N., Hanaoka, S. (2020). Multi-modal relief distribution model for disaster response operations. *Progress in Disaster Science, 6,* 100095.

Ortuño, M. T., Tirado, G., & Vitoriano, B. (2011).A lexicographical goal programming based decision support system for logistics of Humanitarian. *Aid, 19,* 464–479. https://doi.org/10.1007/s11750-010-0138-8.

Sahebjamnia, N., Torabi, S. A., & Mansouri, S. A. (2017). A hybrid decision support system for managing humanitarian relief chains. *Decision Support Systems, 95,* 12–26.

Sheu, J. B. (2007). An emergency logistics distribution approach for quick response to urgent relief demand in disasters. *Transportation Research Part E: Logistics and Transportation Review, 43*(6), 687–709.

Tzeng, G. H., Chiang, C. H., & Li, C. W. (2007). Evaluating intertwined effects in e-learning programs: a novel hybrid MCDM model based on factor analysis and DEMATEL. *Expert systems with Applications, 32*(4), 1028–1044.

Visalakshmi, S., Lakshmi, P., Shama, M. S., et al. (2015). An integrated fuzzy DEMATEL-TOPSIS approach for financial performance evaluation of GREENEX industries. *International Journal of Operational Research, Inderscience Enterprises Ltd, 23*(3), 340–362.

Chapter 7
Analysis of Barriers for the Build the Resilient Supply Chain Networks Post-COVID-19

Krishna Kumar Dadsena, K. Mathiyazhagan, and Atour Taghipour

Abstract Achieving a resilient supply chains network has become a matter of common interest to current supply chain practitioners. A stable, efficient, and resilient supply chain network has the power to change the economical as well as political scenario of any country. Building resilience supply chain networks are often based on the popular discussion as situations where accuracy in decision making helps to improve the operations and to bring back the social activity to normal condition after the pandemic. Hence, a resilient supply chain network affords the probable solutions to handle the associated issues in the network. This study identified and prioritize the barriers in building a resilient supply chain network in India. This research has focused on the situations where supply chain managers are facing challenge's in handling and mitigating different risks by adopting the most suitable strategies. Hence, this challenge has drawn the attention of the current supply chain pre-tensioners to bounce back their organisation from any disruption. In this study, we attempt to provide the impact of COVID-19 on socioeconomic systems and opinions of supply chain managers and provide the recommendations to recover from the current COVID-19 situation and to build their supply chain network more resilient for future challenges. Analytic hierarchy process (AHP) has been adopted for this study. Base on the literature, survey, discussion with the supply chain managers, we have identified factors. Subsequently, based on these factors, a questionnaire was designed and the survey was conducted among supply chain managers. This study provides strategic decision support to managers by considering the post-COVID-19 issues and their effect on their supply chain network. This study also helps managers to provide structural thinking to improve the intuition against the similar scenario in the operational processes that can be envisioned.

K. K. Dadsena
IBS, Shankarapalli Road, Hyderabad 501203, Telangana, India

K. Mathiyazhagan (✉)
Operations & Quantitative Method, Thiagarajar School of Management, Madurai 625015, Tamilnadu, India
e-mail: madii1984@yahoo.com

A. Taghipour
Faculty of International Business, Normandy University, Normandy, France

A. R. Sakthivel et al. (eds.), *Managing Supply Chain Risk and Disruptions: Post COVID-19*, Management and Industrial Engineering,
https://doi.org/10.1007/978-3-030-72575-4_7

Keywords Resilient supply chain networks · Barriers · COVID-19 · Analytic hierarchy process · India

7.1 Introduction

A highly efficient supply chain management is essential to meet the challenges of globalistion and business competition. Supply chain management is a set of supply of the materials at the right time with the help of logistic systems so that the material is produced and distributed with maximising the customer service levels. As the modern supply chains are complex and extended beyond the geographical barriers, they are vulnerable to diverse factors which affect the SCN. Hence, managing the SCN can be considered as an essential activity of the supply chain management to reduce the adverse effects of the different barriers in the performance of the supply chain.

Nowadays, most of the supply chain practitioners are inclined to analyse the barriers to manage their business against increasing uncertain issues such as natural disasters, manmade disasters, competition, customer's expectancy levels and changeable situations. Supply chain (SC) can fail in several ways whose impact depends upon the type of event that occurs, in the supply chain network (SCN) such as logistics network (inbound and outbound network) or sourcing environment (Bode and Macdonald 2017). As the impacts of COVID-19 are being felt by the practitioners day by day, despite efforts by the individual, government to control its effects, which is triggering the loss of human life and business across the global market. The COVID-19 pandemic is an interruption in almost every activity of a business, which affects the supply chains network adversely. "Nearly 75 percent of companies testified supply chain interruption due to COVID-19 pandemic and it is expected that this figure will rise over the time in future" according to Institute For Supply Chain Management (Sengupta 2020). Therefore, understanding the impacts of COVID-19 in the SCN has become an important challenge to make the network more resilient in the future.

An efficient and resilient SC is the demand of the time to sustain in the global dynamic market economy as well as social growth of the country. A scientific and systematic approach is required to first analyse the barriers in building the SC resilient against a negative event. The current SC practitioners' emphasis more on the need for a structured understanding and analysing their network for goal setting, allocation and prioritisation of resources to supports the managers in either approving or changing current policy considering the COVID-19 pandemic.

Due to the COVID-19 pandemic making supply chain resilience has become one of the main subjects for SC practitioners, consultancies, insurance companies, government agencies and international organisations. Hence, to sustain against this pandemic which leads to barriers in the SC management process. Ali et al. (2017) advocated the need for analysing the barriers in SC management and considered it as an integral part of the risk management process. This helps in improving the performance by understanding the uncertain activities and also helps in the development of mitigation strategies. Considering the current COVID-19 pandemic there is still extensive study that needs to be conducted with regards to the selection of appropriate barriers after the pandemic.

7.2 Literature Review

The term resilience has been discussed in different fields like ecology, metallurgy, psychology, and management (Ali et al. 2017). In the supply chain perspective, the context of the supply chain, the most widely accepted and theoretically (Ponomarov and Holcomb 2009) defined the supply chain resilience as the capability of SCN to prepare, respond and recover from the unexpected disruption in the operation of SC. The factors that influence the SCR are such as SC visibility inventory availability, strategy planning and implementation (Macdonald et al. 2018). Hence, identification and selection of the barriers in the SCR playas an important role in such situations. Undoubtedly, the pandemic COVID-19 is an unexpected and extraordinary event facing the global SC practitioners with a question that continued functioning of SCN with the current SCN design (Ivanov and Das 2020). Due to the COVID-19 outburst the academicians and practitioners are struggling to manage their SCN in several perspectives such as there may be loss of opportunities because of improper communication and lack of proper information about the load for the transporter (Ivanov 2020). Considering the size of the SCN and the increasing difficulty of its operations, it becomes important to study and identify the various SCR factors, prioritize them based on their criticality.

Due to COVID-19 pandemic, an effective and efficient design of the SCN has become a big challenge for practitioners (Kumar et al. 2020). Authors advocate the need for a systematic and scientific approach to manage against the current challenges status and future trend of evolving vulnerabilities in the operational process. Golan et al. (2020) highlighted the impact of the COVID-19 pandemic on global and local SCN and emphasizes an advanced resilience analytical approach that can help the managers in the strategic decision-making process of SCN. Therefore, the accurate, timely and convincing decision considering the uncertainty in the operations has become one of the main challenges in such situations. This has motivated us to take up the challenge of the identification of different barriers in SCN. Thereby, this study aims to fill this research gap by extending the approach for the selection of barriers in resilience SCN design post-COVID-19.

Due to complex and uncertain operating environment arises in SCN due to COVID-19, almost all type of industry is facing several risky situations that might lead to overall operational failure of their SCN. Sharma et al. (2020) described that the risks involved in the individual activity of the supply chain contribute to the complete failure of the whole supply chain network. Golan et al. (2020) recommended that identification and reduction of operational uncertainty due to the COVID-19 pandemic helps the SCN to improve its resilience against a similar situation in near future.

Considering the intense situation like COVID-19 the multi-criteria decision-making approach has been proven its effectiveness in decision making Sharma et al. (2020). Also, recent articles by authors, Albahri et al. (2020), Belhadi et al. (2020), Yang et al. (2020) emphasises the use of a multi-criteria decision-making approach to help management in decision making in a pandemic situation. The study by Mahato et al. (2020) and Pamungkas et al. (2020) has been effectively applied the AHP in decision making under the pandemic situation.

7.3 Solution Methodology

AHP methodology has been adopted as a solution tool for this study shown in Fig. 7.1. It helps in decomposing, organising, and analysing a complicated decision-making problem by considering different criteria and sub-criteria (Luthra et al. 2016). This approach converts the problem into a hierarchical structure category in different levels, such as main objective or goal, criteria, and sub-criteria (Govindan 2015; Mangla et al. 2015). Therefore, AHP is recommended as a better tool compares to others due to its extensive applicability and user-friendly nature (Harputlugil et al. 2011). Thus, in this study, we have used the AHP approach to evaluate barriers in resilience SCN for Post COVID-19 situations (Tables 7.1 and 7.2).

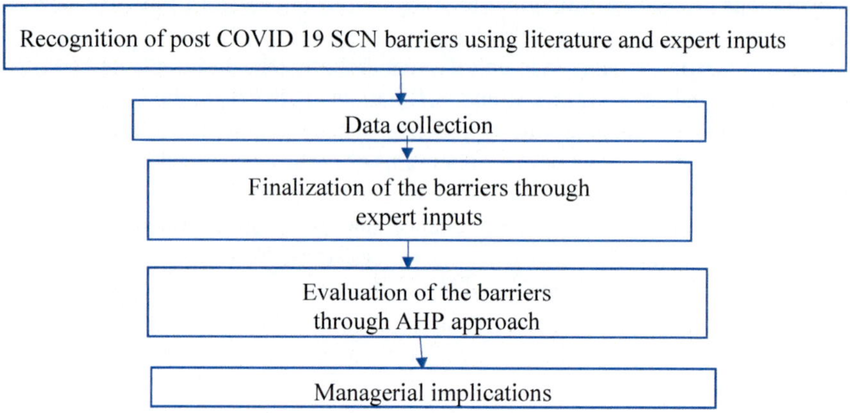

Fig. 7.1 Steps of the proposed research

Table 7.1 The resilient supply chain networks post COVID-19 criteria

Main criteria	Sub-criteria	Description	Source
Manufacturing (D1)	The traditional centralised manufacturing model (D11)	Industries will have to reduce risk against dependencies on centralized manufacturing	Pearce (2020), Mandal (2020)
	Intelligent and responsive allocation across channels (D12)	Businesses practitioners should pay attention to make their SCN more capable towards the intelligent and responsive system to improve the effectiveness of their business	Ivanov (2020), Sengupta (2020)
	Little understanding and flexibility of inventory management system (D13)	Managing Inventory requires more focus by the planners as they may face the challenge	Patrinley et al. (2020), De Kock (2020)
Operational barriers (D2)	Inability resume operation quickly (D21)	About resuming operations and getting back to the normal operating state	DePhillipo (2020)
	Deficiency in risk identification and wrong assumptions (D22)	The knowledge and collective action to mitigate the risk can help in quick departure from the disaster possible	Golan (2020), McMaster et al. (2020)
	Failure in quick operational changes for challenges (D23)	The Supply Chain Managers requires a flexible and efficient SCN through enhancing the agility in the operation	Baghalian et al. (2013)
Sourcing and Distribution Barriers (D3)	Capacity or inventory inflexibilities (D31)	Examining and improving the capacity or inventory management system	McMaster et al. (2020)
	Sudden changes in demand (Bullwhips effect) (D32)	Managers should focus on uncertainty on supply and demand	Patrinley et al. (2020)

(continued)

Table 7.1 (continued)

Main criteria	Sub-criteria	Description	Source
	Need for direct distribution (D33)	The change in distribution design may help to make SCN resilient	Baghalian et al. (2013)
Human and judgmental barriers (D4)	Complying with service level agreements (D41)	Pandemic sharpens the managers thinking about strengthening the service quality	Sharma et al. (2020)
	Lack of experience in dealing with such risks (D42)	In aiming for SCN resilient managers need to identify and manage the negative events effectively	McMaster et al. (2020)
	Lack of expanding the individual responsibility (D43)	The resilient SCN needs individuals to response	Pan and Zhang (2020)

Table 7.2 Numerical ratings in the AHP method

Scale	Meaning (A factor compared to B)
1	Equal importance
3	Moderate importance
5	Strong importance
7	Very strong or demonstrated importance
9	Extreme importance
2, 4, 6, 8	Intermediate values
Reciprocals	When B is compared with A

7.3.1 Step 1: Hierarchal Framework

See (Fig. 7.2).

7.3.2 Step 2: Develop the Pairwise Comparison Matrix

Pairwise comparison matrix of main criteria

	D1	D2	D3	D4
D1	1	0.111	0.142	0.200
D2	9	1	3	5

(continued)

(continued)

	D1	D2	D3	D4
D3	7	0.333	1	2
D4	5	0.200	0.500	1
Sum	22	1.644	4.642	8.200

Pair wise comparison for the sub-criteria for Manufacturing criteria				Pair wise comparison for the sub-criteria for Operational barriers			
	D21	D22	D23		D21	D22	D23
D11	1	0.1428	0.2	D21	1	0.1111	0.25
D12	7	1	3	D22	9	1	3
D13	5	0.3333	1	D23	4	0.3333	1
Sum	13	1.4761	4.2	Sum	14	1.4444	4.25
Pair wise comparison for the sub-criteria for Sourcing and Distribution Barriers				Pair wise comparison for the sub-criteria for Human and judgmental barriers			
	D31	D32	D33		D41	D42	D43
D31	1	0.1428	0.3333	D41	1	0.1111	0.2
D32	7	1	2	D42	9	1	7
D33	3	0.5	1	D43	5	0.1428	1
Sum	11	1.6428	3.3333	Sum	15	1.2533	8.2

7.3.3 *Step 3: Weights of the Main and Sub-Criteria*

See (Table 7.3).

7.4 Managerial Implications

This study helps SC practitioners to become more informed and resilient against barriers in post-COVID-19 pandemics. As it can be seen from Table 7.3 operational barriers secured the highest weight this is indicative of the fact that the operation of SC would be challenging for the practitioners. Also, the sub-criteria show that managers should focus on the systematic and scientific approach to handle the risks carefully. This result will help the managers to know the factors on which he should concentrate on the efficient operation of his business. The article also provides the decision-making information which will enable the manager in implementing the appropriate strategy to control the situation. This study supports management to prepare well for the post-COVID-19 pandemic. It is found that specific call to risk

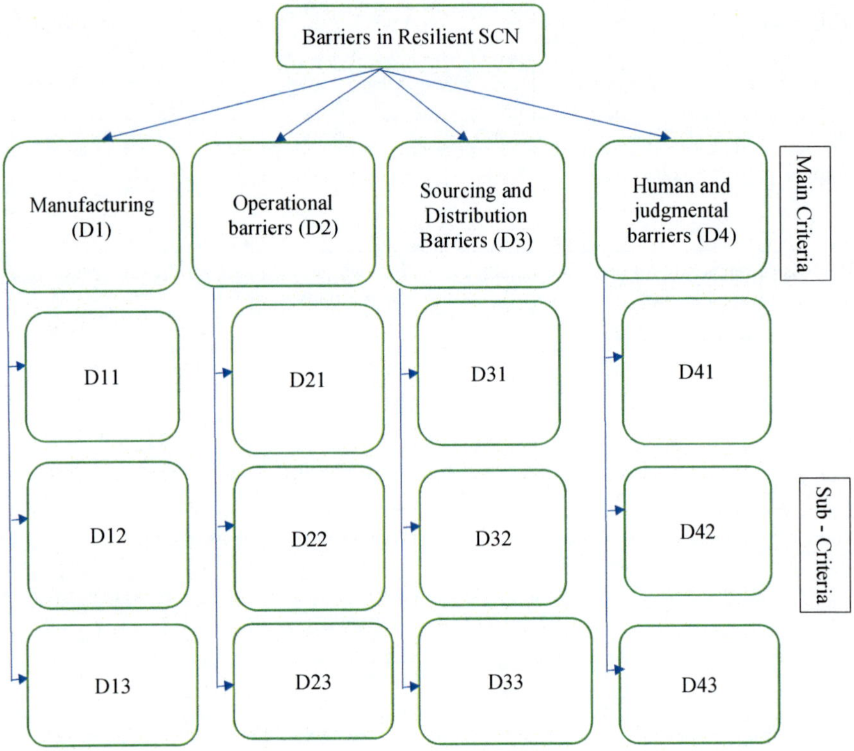

Fig. 7.2 Main criteria and sub-criteria

Table 7.3 Weights of the main and sub-criteria

Main Criteria	Criteria Weight	Sub-criteria	Criteria Weight	Global Weight
Manufacturing (D1)	0.0420	D11	0.0737	0.0031
		D12	0.6433	0.0270
		D13	0.28283	0.01189
Operational barriers (D2)	0.5682	D21	0.0690	0.0392
		D22	0.6803	0.3866
		D23	0.2505	0.1424
Sourcing and distribution barriers (D3)	0.2450	D31	0.0926	0.0226
		D32	0.6150	0.1507
		D33	0.2923	0.0716
Human and judgmental barriers (D4)	0.1446	D41	0.0598	0.0086
		D42	0.7503	0.1085
		D43	0.1897	0.0274

managers to recognize and adopt responsive action throughout the channel to reduce risk by avoiding the wrong assumption and improve their risk management process.

7.5 Conclusion

To sustain in the rapidly growing competitive global market, due to COVID-19 pandemic, SC practitioners need to identify the barriers and its management. Which helps the organisation to take a better-informed decision with an improved probability of achieving its strategic and operational objectives. This study presents the barriers in the SCN design to sustain post-COVID-19 situations, as the complexity and uncertainty involved in the operational process became a big challenge to the SC managers. AHP has been proven as an effective and most widely used decision-making tool to provide the weightage for the different criteria based on their relative importance.

This work identified and evaluated the barriers associated with the post-COVID-19 pandemic in the SCN. Initially, four main categories of barriers and twelve sub-criteria related to post-pandemic have been identified from the trade journal, magazine, literature. Then, AHP has been used to evaluate and rank these barriers by calculating the weightage of the criteria. The use of the proposed approach directly influences the managers to take an operational decision in the performance improvement of the industry. SCN encompasses different attributes, the proposed approach provides an aggregated opinion of the different stakeholders to reach a common conclusion in finding the barriers and their mitigation strategy.

In the future, this study can be extended in many directions. In future research, fuzzy AHP may be used to eliminate ambiguity and uncertainty. Also, the development and selection of strategy considering economic constraint for the SC manager will be one of the very interesting future works.

References

Albahri, O. S., Al-Obaidi, J. R., Zaidan, A. A., Albahri, A. S., Zaidan, B. B., Salih, M. M., et al. (2020). Helping doctors hasten COVID-19 treatment: Towards a rescue framework for the transfusion of best convalescent plasma to the most critical patients based on biological requirements via ml and novel MCDM methods. *Computer Methods and Programs in Biomedicine, 196,* 105617.

Ali, I., Nagalingam, S., & Gurd, B. (2017). Building resilience in SMEs ofperishable product supply chains: enablers, barriers and risks. *Production Planning & Control.* https://doi.org/10.1080/09537287.2017.1362487.

Baghalian, A., Rezapour, S., & Farahani, R. Z. (2013). Robust supply chain network design with service level against disruptions and demand uncertainties: A real-life case. *European Journal of Operational Research, 227*(1), 199–215.

Belhadi, A., Kamble, S. S., Khan, S. A. R., Touriki, F. E., & Kumar, D. (2020). Infectious waste management strategy during COVID-19 pandemic in africa: An integrated decision-making framework for selecting sustainable technologies. *Environmental Management*, 1–20.

Bode, C., Macdonald, J. R. (2017). Stages of supply chain disruption response: Direct, constraining, and mediating factors for impact mitigation: stages of supply chain disruption response. *Decision Sciences, 48*(5), 836–874. https://doi.org/10.1111/deci.12245.

De Kock C. (2020). The perplexities for inventory planners post COVID-19.

DePhillipo, N. N., Larson, C. M., O'Neill, O. R., & LaPrade, R. F. (2020). Guidelines for ambulatory surgery centers for the care of surgically necessary/time-sensitive orthopaedic cases during the COVID-19 pandemic. *The Journal of Bone and Joint Surgery*. American Volume.

Golan, M. S., Jernegan, L. H., & Linkov, I. (2020). Trends and applications of resilience analytics in supply chain modeling: systematic literature review in the context of the COVID-19 pandemic. *Environment Systems & Decisions, 1*.

Govindan, K., Khodaverdi, R., & Vafadarnikjoo, A. (2015). Intuitionistic fuzzy based DEMATEL method for developing green practices and performances in a green supply chain. *Expert Systems with Applications, 42*(20), 7207–7220.

Harputlugil, T., Prins, M., Gultekin, T., Topcu, I. (2011). Conceptual framework for potential implementations of multi criteria decision making (MCDM) methods for design quality assessment. in *Proceedings of the Management and Innovation for a Sustainable Built Environment*, Amsterdam, The Netherlands June 20–23. ISBN: 9789052693958. https://hbr.org/2020/08/leading-into-the-post-covid-recovery.

Ivanov, D. (2020). Predicting the impacts of epidemic outbreaks on global supply chains: A simulation-based analysis on the coronavirus outbreak (COVID-19/SARS-CoV-2) case. *Transportation Research Part E: Logistics and Transportation Review, 136*, 101922.

Ivanov, D., & Das, A. (2020). Coronavirus (COVID-19/SARS-CoV-2) and supply chain resilience: A research note. *International Journal of Integrated Supply Management, 13*(1), 90–102.

Kumar, M. S., Raut, R. D., Narwane, V. S., & Narkhede, B. E. (2020). Applications of industry 4.0 to overcome the COVID-19 operational challenges. *Diabetes & Metabolic Syndrome: Clinical Research & Reviews*.

Luthra, S., Mangla, S. K., Xu, L., & Diabat, A. (2016). Using AHP to evaluate barriers in adopting sustainable consumption and production initiatives in a supply chain. *International Journal of Production Economics, 181*, 342–349.

Macdonald, J. R., Zobel, C. W., Melnyk, S. A., & Griffis, S. E. (2018). Supply chain risk and resilience: Theory building through structured experiments and simulation. *International Journal of Production Research, 56*(12), 4337–4355.

Mahato, R., Nimasow, G., Dai Nimasow, O., & Bushi, D. (2020). Analytic hierarchy process based potential risk zonation of COVID-19 in India. *Journal of the Social Sciences, 48*(3).

Mandal, D. (2020). Coronavirus boost for Industry 4.0: From cyber preparedness to skill building, how businesses must adapt. https://www.financialexpress.com/industry/coronavirus-boost-for-industry-4-0-from-cyber-preparedness-to-skill-building-how-businesses-must-adapt/2008641/.

Mangla, S. K., Kumar, P., Barua, M. K. (2015). Risk analysis in green supply chain using fuzzy AHP approach: a case study. *Resources, Conservation and Recycling, 104*(B), 375–390.

McMaster, M., Nettleton, C., Tom, C., Xu, B., Cao, C., & Qiao, P. (2020). Risk management: Rethinking fashion supply chain management for multinational corporations in light of the covid-19 outbreak. *Journal of Risk and Financial Management, 13*(8), 173.

Pamungkas, T. S., Nugroho, A. S., Wasiso, I., Anggoro, T., & Kusrini, K. (2020). Decision support system for direct target cash recipients using the AHP and K-Means method. *Research: Journal of Computer, Information System & Technology Management, 3*(2), 45–54.

Pan, S. L., & Zhang, S. (2020). From fighting COVID-19 pandemic to tackling sustainable development goals: An opportunity for responsible information systems research.

Patrinley, J. R., Berkowitz, S. T., Zakria, D., Totten, D. J., Kurtulus, M., & Drolet, B. C. (2020). Lessons from operations management to combat the COVID-19 pandemic. *Journal of Medical Systems, 44*(7), 1–2.

Pearce, J. M. (2020). Distributed manufacturing of open source medical hardware for pandemics. *Journal of Manufacturing and Materials Processing, 4*(2), 49.

Ponomarov, S. Y., & Holcomb, M. C. (2009). Understanding the concept of supply chain resilience. *The International Journal of Logistics Management, 20*(1), 124–143.

Sengupta, A. (2020). Impact of COVID-19 on global supply chains and opportunities in the post-COVID world. www.entrepreneur.com/article/349229. Accessed 20 July 2020.

Sharma, M., Luthra, S., Joshi, S., & Kumar, A. (2020). Developing a framework for enhancing survivability of sustainable supply chains during and post-COVID-19 pandemic. *International Journal of Logistics Research and Applications*, 1–21.

Yang, Z., Li, X., Garg, H., & Qi, M. (2020). Decision support algorithm for selecting an antivirus mask over COVID-19 pandemic under spherical normal fuzzy environment. *International Journal of Environmental Research and Public Health, 17*(10), 3407.

Chapter 8
Action Plans for Logistics and Supply Chain Recovery Post-COVID-19

Deepak Mathivathanan and K. Sivakumar

Abstract The unprecedented humanitarian crisis caused due to the COVID-19 breakout has irrevocably changed the global economic landscape and has posed huge challenges to business owners around the globe. Global supply chains are further affected due to the government policies on travel restrictions, quarantines and suspending business operations intended to prevent the further spread of the virus. The evolution of various supply chain strategies over the years focused on supply chain optimization to minimize costs, inventories and increase effective utilization of resources has forced firms to go for minimized buffers and capability to absorb vast scale disruption caused by the COVID-19 pandemic. Significant logistics and supply chain challenges will remain for an indeterminate period of time as pandemic impacts still continue to peak in developing countries. Further, with the world entering a new normal post-COVID-19, drastic changes in overall spending and consumer behaviour are. Though many businesses are adapting crisis-management mechanisms their focus is typically on the short term. However, logistics and supply chain planners hope to plan the medium and long term to build resilience to see them through the other side. Hence, this chapter focuses on the set of issues that require immediate attention across the entire supply chains amidst the ongoing pandemic and further we discuss the action plans to mitigate the pandemic effect and to recover by building resilience against such future shocks.

Keywords COVID-19 pandemic · Digital supply chain · Recovery · Action plan · Resilience

D. Mathivathanan (✉)
Centre for Logistics and Supply Chain Management, Loyola Institute of Business Administration, Chennai, Tamil Nadu, India
e-mail: deepak.mathivathanan@liba.edu

K. Sivakumar
Department of Mechanical Engineering, Siddharth Institute of Engineering and Technology (SIETK), Puttur 517583, Andhra Pradesh, India

A. R. Sakthivel et al. (eds.), *Managing Supply Chain Risk and Disruptions: Post COVID-19*, Management and Industrial Engineering,
https://doi.org/10.1007/978-3-030-72575-4_8

8.1 Introduction

The COVID-19 disease which began, has a health scare, in Wuhan, China has transformed into a global pandemic spread across more than 200 countries and already disrupted the global economic landscape irrevocably. Governments across the globe declaring either partial or complete lockdown measures as a strategy to control the spread of COVID-19 has led to massive disruptions to international trade and global supply chains. As a result, global GDP has taken a serious hit, much faster and worse than the economic crisis of 2008. With automotive, textiles, aviation, tourism and electronics sectors being worst hit by the pandemic (Chamola et al. 2020) the sectors are focusing on a quick rebound, the resumption of business as usual is yet far away in time.

Severe disruptions in inbound and outbound logistics like air cargo shipments, trucking and rail cargo services, heavy port congestions are witnessed because of the coronavirus crisis (Linton and Vakil 2020). Further, the regional lockdowns have impeded logistics operations causing unusual delays in shipments. Suddenly, the businesses realized their supply chains were no more resilient and consumers who were used to abundance found themselves with scarcity issues even in their daily basic needs (Ivanov and Das 2020). Post-COVID-19 the logistics and supply chain management will not be the same as normal. Businesses will have to undergo significant transitions in the way they operate to adjust to the new normal. Adjusting to the new demand scenarios, transportation options, efficient warehousing & operations and building resilience in the logistics and supply chain networks are the key issues to be addressed. The pre-COVID supply chains must build resilience by going for multiple suppliers and diversified manufacturers. The supply chain priorities must be re-assessed based on the current realities and the associated warehousing and logistics might have to be re-configured to be more flexible and yet cost-effective. Supply chain innovations both technological and operational along the entire value chain will be the way to move forward post-COVID-19. Thus, the impact of coronavirus has been significant on logistics and supply chain management around the globe and in future the supply chain planners and professionals must focus on supply chain transparency and continuous monitoring.

Several businesses worldwide have mobilized and initiated rapid crisis management to ensure their business continuity in the short-term. But, to achieve logistics and supply chain resilience medium- and long-term planning is required to recover from the impact created by the COVID-19 outbreak. Examining closely, the need for a resilient supply chain and logistics network through innovative recovery approaches is the need of the hour (Blackhurst et al. 2011; Hosseini et al. 2019; Remko 2020; Singh et al. 2020). Planning for supply chain recovery involves keeping the logistics system agile to ensuring continuous ship movements, keeping the ports open for international trade flows and importantly not compromising on ensuring the safety measures to control the spread of the COVID-19 virus (UNCTAD 2020).

The logistics and supply chain recovery process requires a complete reassessment of the existing infrastructures in terms of location, production capabilities, capacities, management of information flows, material flows to effectively reconfigure the existing structure. Integration and collaboration among partners will be crucial to establish common goals and share not only the future gains but also the losses incurred. Knowledge management and data gathering within supply chain members will play important role in identifying and understanding bottlenecks to provide innovative solutions. After addressing the disruptions, logistics and supply chain managers should work on ensuring trust and transparency across the supply chain. Critical supply chains such as pharma and food supply chains would require the most attention from policymakers and government in general aftermath of COVID-19. Consumption patterns of customers and changes in consumer behaviours are witnessed along with the increased usage of social media and chatting apps used for sharing items. Sustainable consumption, buying local and ethical consumer goods are increasing along with the use of service provider apps. Hence, studies on logistic and supply chain recovery post-COVID-19 have gained paramount interest with both academicians and industrial experts. This chapter intends to address this need for the possible action plans for post-COVID-19 logistics and supply chain recovery in two main exclusive themes, that is, logistics recovery and supply chain recovery.

8.2 Logistics Recovery Post-COVID-19

The logistics industry has always been the key element of global trade and economy by ensuring just-in-time delivery to reduce inventory cost, driving global market penetration, enabling accurate and consistent demand forecasting and associated planning, delivering low-cost sourcing and eventually increasing customer satisfaction by speeding the deliveries. More importantly, the quarantine policy has led to the supply decrease of drivers and trucks (Cai and Luo 2020). But, as a part of reducing the international spread of the COVID-19 virus, policymakers from most countries need to ensure safer international logistics and transport to keep the trade going on. Implementing drastic safety measures and controls against COVID-19 and the need to keep the cross-border trade flowing are of paramount importance. A decrease in production and lower demand for products barring the essentials have reduced the shipping volumes across the globe. Many containers firms are at the brink of bankruptcy and many firms are faced with unforeseen struggles to collect and analyze data for necessary risk mitigation strategies. With lower production scale and varying demand scenarios suppliers are unable to forecast effectively creating a bullwhip effect in distribution trade. Further, with the airline business taking a serious hit, lack of resources is expected and on the other hand, the increase in online shopping and panic buying the existing logistics infrastructure prove to be insufficient.

Though the signs of returning to normalcy are far away in time serious action plans must be followed to ensure that the logistics industry comes out of this COVID-19 crisis. A prioritized restoration of supply chains primarily meeting the essential product demands and gradually bring the demand for other products to normal will prevent overburdening of at present limited supply system. Creating efficient contingency plans to avoid delays and bottlenecks becomes a key element in tackling unexpected circumstances. Training the employees on hygiene and social distancing will reflect in gaining consumer confidence. Taping the technological advancements in artificial intelligence (AI), machine learning (ML) to create real-time supply chain dashboards, Global Positioning System (GPS) tracking is going to be the way to go. This section summarizes the concrete measures proposed by United Nations to facilitate safe transport and trade across the border and eventually protecting people from contracting the virus.

- Since the maximum of global trade is via commercial shipping it is necessary to provide services from bunkering to supplies without compromising on the health regulations and ensure uninterrupted shipping around the clock.
- Keeping the ports open to ships and intermodal connections 24/7 to spread workloads to protect port personnel and to maintain efficient port operations.
- Speeding up custom clearance and protecting international trade in cases of critical goods by providing a harmonised system for critical medical supplies to facilitate accelerated and fast-track clearances.
- Facilitating cross-border air freights for urgent deliveries, enabling trucks, trains and transport workers to cross borders through pre-processing of clearances and approving relevant storage facilities are critical.
- Transit countries co-operating and maintaining their access to seaports and other trade corridors to ensure a steady flow of goods in landlocked countries.
- Governments supporting and assisting industrial associations to share information and enabling transparency.
- Going for paperless transactions by adopting electronic alternatives to minimize the risk of COVID-19 transmission via contaminated surfaces can protect the health of workers.
- Promoting collaborative action plans to counter the legal implications of delays, performance drop and other related issues need to put into effect wherever appropriate.
- For the benefit of vulnerable countries providing financial support will ensure resilience in supply chains and enable them to stay in business.
- Providing tailor-made advisory services, research programs to enable capacity building, trade and customs automation for all supply chain partners.

8.3 Supply Chain Recovery Post-COVID-19

In comparison to logistics, the supply chain recovery is much complex. Organizations are required to continuously monitor both short- and long-term demand to manage inventory and accommodate production losses incurred due to continued factory shutdowns. Also, on the other hand, retailers face a problem of panic buying which leaves them with a depleted inventory. Further, customers are now more comfortable with home delivery options leaving retailers with the additional burden of managing increased last-mile delivery cases while also keeping their employees safe. Creating contingency plans to face the unique risks each company faces becomes more and more important. Planning for post-COVID, companies should plan for both optimistic and conservative scenarios because returning to normalcy is uncertain. Strategies to mitigate supply shocks can be dealt successfully by planning effectively with existing suppliers. Leveraging the aggregated data by third-party partners enables increased visibility of the movement of goods across the network and enables the organizations to frame effective contingency plans. With limited supplies, managing demand volatility without compromising on worker safety is of key importance. Workforces handling food products, or any other last-mile deliveries should be educated about their and customer safety. Also, checking their health regularly and providing them with protective equipment will guard them against COVID-19. Following this outbreak, factories and firms are moving towards reopening their business. Though COVID-19 may be an unpredictable black swan event, understanding one's own supply chain in terms of its risk, resilience and opportunities can enable firms to plan better for the crisis.

Post-COVID-19 disruption, manufacturing firms and even retailers will look beyond the cheapest places for safe and best places to the source, produce and for storing the products. Planning COVID-19 task forces at executive and operational levels to address challenges and partnering with suppliers to enable critical shipments to flow is important. Many factories have opted for dividing their workforces into shift-based groups and furthermore, companies have gone to the extent of harnessing the AI-driven manufacturing alternatives for the existing capacity. In the case of grocery supplies, sourcing and procurement teams should investigate digital solutions to switch on to additional sources and readjust to having flexible sources to need varying demands. Post-COVID-19 stock allocation decisions will more be taken at the distributor level. In general, the companies which have implemented supply chain risk management strategies will be able to ensure business continuity. Multi-sourcing and buffering strategies have come to their rescue. Firms with strong supplier relationships to leverage visibility across supply chain networks developed the agility to reconfigure their distribution networks. On the other hand, the firms reliant on a single supplier strategy and those who do not have end-to-end visibility failed to understand the current inventory status to ensure the uninterrupted flow of products. Though, adopting Digital supply network models as shown in Fig. 8.1 could solve the current challenges many short-term actions in line with achieving long-term goals are needed to respond to current challenges and manage cash flow and net working capital immediately.

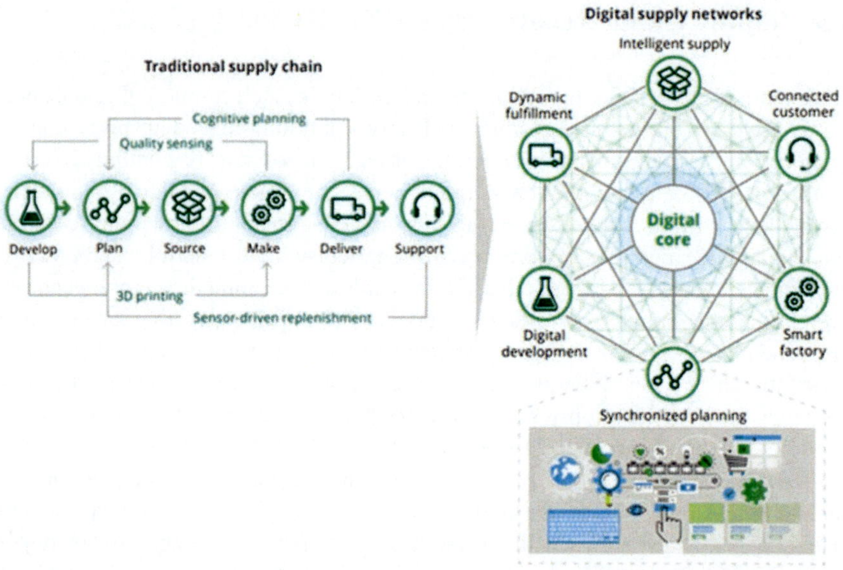

Fig. 8.1 Transforming traditional supply chains to digital supply networks. Adopted from Gaus et al. (2018)

Creating transparency: Working closely with the production and operations team will help in identifying the components or services outsourced from high-risk suppliers and assess the risk of supply interruptions. For cases with the risks of significantly interrupting the production schedules, alternate suppliers need to be identified or firms can collaborate with their suppliers to address the risks. Understanding such specific risks across the supply chain will allow a faster restart post-COVID-19 and this could be made possible only by creating transparency across the supply chain.

Inventory management: Built-in inventory in supply chains can postpone the impact of halted production lines and buy time to plan effective recovery strategies. Transparency in supply chains can aid in estimating inventory across the value chains. All categories namely finished goods, spare parts, parts in transit, supplies with dealers need to be estimated to ramp up post-COVID-19 recovery.

Optimizing production and distribution capacity: Pandemic crisis like COVID-19 has significantly increased or decreased the demand for different products. Using statistical forecasting tools and ensuring continuous monitoring of forecasts can enable companies to optimize their production and distribution capacities accordingly. To ensure employee safety educating them about infection risks and enabling work at home options can be tailor-made by assessing firm-specific scenarios.

Securing logistics capacity: As discussed in the earlier section, COVID-19 has disrupted any possible logistics across the globe. Hence, it is critical to prioritize the logistics needs and prebook logistics facilities to minimize the potential cost increases. Collaborating with partners can effectively improve the chances of securing logistics capacity. Also, changing the mode of transport based on given scenarios and accelerating customs clearance are other useful aids.

Assess supply chain vulnerability: Finally, the potential drivers of supply chain vulnerability should be addressed. Firms should look upon addressing issues like, how predictable is demand planning? how exposed is their network to disruptions? How resilient is their transportation network? How vulnerable is their product to regulatory changes? and how could they mitigate future disruptions? By acting effectively over the next few months on the above issues, companies can better their chances of recovering post-COVID-19 as soon as possible.

8.4 Need for Digital Supply Chain Model for Post-COVID-19 Recovery

Supply chain sophistication has increased exponentially over the recent years and it has been the major factor in building any firms' competitive edge. Over the years, supply chain optimizations to minimize costs and inventories have hindered the flexibility to accommodate disruptions as caused by COVID-19. The traditional supply chain models and risk management techniques may not absorb global shocks effectively. Fortunately, innovative digital technologies present today can enhance end-to-end visibility, agility, and responsiveness to mitigate any unforeseen disruptions. Transformation of the linear supply chain into digital supply networks aligned with overall business strategy will build a resilient supply chain that not only reduces the risks of any disruptions but also enable firms to recover faster. With the fourth industrial revolution (Industry 4.0) gaining much attention, technologies such as IoTs, AI, Cloud computing, Robotics, Cyber-Physical systems, Blockchain technology, Big Data analysis, Additive manufacturing are worth investing to enable an effective digital supply network for the future.

8.5 Conclusion

This chapter presented the corresponding actions to be taken in order to maintain smooth logistics operations and enhance the resilience of the supply chain post the COVID-19 pandemic. There will be profound changes to the economy and social life after the pandemic. By supporting supply chain partners and maximizing the usage of favourable policies implemented within the period, the supply chain operation is expected to be gradually recovered to some extent. Similarly, ensuring employees' safety, continued smooth functioning of cross-border transport, protecting the shippers and transport service providers and speed up customs clearance and trade facilitation during the pandemic can improve the delivery of essential goods to the affected zone. With the implementation of these measures in many industries and logistics sectors globally, the average recovery rate of logistics and supply chain for industrial enterprises above designated size. Implementing the suggested measures post the pandemic will assist the managers in the ease of logistic trade and resilience of the supply chain. The action plans considered in this chapter are generic, and thus, they can be used to extend the study with insertion or elimination of a few more measures. Furthermore, this chapter is a useful source to build the resilience supply chain and logistics to cope with any other pandemic outbreak by any organization.

References

Blackhurst, J., Dunn, K. S., & Craighead, C. W. (2011). An empirically derived framework of global supply resiliency. *Journal of business logistics, 32*(4), 374–391.

Cai, M., & Luo, J. (2020). Influence of COVID-19 on manufacturing industry and corresponding countermeasures from supply chain perspective. *Journal of Shanghai Jiaotong University (Science), 25*(4), 409–416.

Chamola, V., Hassija, V., Gupta, V., & Guizani, M. (2020). A comprehensive review of the COVID-19 pandemic and the role of IoT, Drones, AI, blockchain, and 5G in managing its Impact. *IEEE Access, 8*, 90225–90265.

Gaus, T., Olsen, K., & Deloso, M. (2018). Synchronizing the digital supply network. *Deloitte Insights*.

Hosseini, S., Ivanov, D., & Dolgui, A. (2019). Review of quantitative methods for supply chain resilience analysis. *Transportation Research Part E: Logistics and Transportation Review, 125*, 285–307.

Ivanov, D., & Das, A. (2020). Coronavirus (COVID-19/SARS-CoV-2) and supply chain resilience: A research note. *International Journal of Integrated Supply Management, 13*(1), 90–102.

Linton, T., & Vakil, B. (2020). Coronavirus is proving we need more resilient supply chains. Harvard business review, 5. https://hbr.org/2020/03/coronavirus-is-proving-that-weneed-more-resilient-supply-chains.

Remko, V. H. (2020). Research opportunities for a more resilient post-COVID-19 supply chain–closing the gap between research findings and industry practice. *International Journal of Operations & Production Management, 40*(4), 341–355.

Singh, S., Kumar, R., Panchal, R., & Tiwari, M. K. (2020). Impact of COVID-19 on logistics systems and disruptions in food supply chain. International Journal of Production Research, 1–16.

UNCTAD. (2020). COVID-19: A 10-Point action plan to strengthen international trade and transport facilitation in times of pandemic. https://unctad.org/en/pages/newsdetails.aspx?OriginalVersionID=2311.

Chapter 9
Need for Embracing New Digital Technologies in Healthcare Supply Chain

Manavalan Ethirajan and Jayakrishna Kandasamy

Abstract The supply chain has become even more important in the prevailing crisis like COVID-19 as the impact is seen worldwide. Any disruption in the supply chain affects the entire economy and the COVID-19 pandemic has proven it one more time especially for the healthcare sector. In the current pandemic situation, the customers (patients) need to vary much due to the diversified requirements of the patient. Thus, the study on digital technologies in healthcare is essential to bring more value the organisations. This study intends to contribute to the health care supply chain to assess the new technologies. Based on the review, a framework for the healthcare supply chain is developed and the key criteria are analysed using Quality Function Deployment (QFD) based approach. The conceptual framework is designed based on three enablers namely inventory, organisation, and customer service in the healthcare supply chain. Based on the analysis, new technologies to fulfill the gap in the present system are proposed. It is encouraging to see the management is ready to deploy technology as the inputs are based on the voice of different stakeholders. This pandemic has shown that the health care supply chain needs to embrace new technologies such as Big Data, Cloud Computing, Artificial Intelligence, Internet of Things to improve overall supply chain efficiency.

Keywords Health care supply chain · Digital technologies · QFD · COVID-19

9.1 Introduction

The COVID-19 pandemic has exposed vulnerabilities in the current healthcare supply chain (Govindan et al. 2020). As healthcare has strived to maintain required inventory in terms of Masks, Personal Protective Equipment, Ventilators, and drugs (Nikolopoulos et al. 2020). Healthcare management has shifted its focus on supply chain optimisation now and prioritising technology-based initiatives to fight against Covid-19 (Leite et al. 2020).

M. Ethirajan · J. Kandasamy (✉)
School of Mechanical Engineering, Vellore Institute of Technology, Vellore 632014, Tamil Nadu, India

A. R. Sakthivel et al. (eds.), *Managing Supply Chain Risk and Disruptions: Post COVID-19*, Management and Industrial Engineering,
https://doi.org/10.1007/978-3-030-72575-4_9

Most of the hospitals are desperate to get the supplies seamlessly and want to avoid stockout on critical medicines and devices (Singh et al. 2020). Some of the hospitals have the data but unsure of how to use it to drive decisions. There are multiple challenges faced by healthcare during this pandemic, some hospitals have placed multiple orders for the same items mistakenly and are not ready to pay the supplier. In addition, there are few organisations that struggled to get the minimum number of healthcare items as some of the suppliers started hoarding supplies of hand sanitiser, personal protective equipment as they have seen a sudden surge in demand for these items. Eventually, there is a shortage of these basic healthcare items to hospitals and frontline workers. Some consumers started panic buying things in excess and stored them in fear. These challenges have made the health care supply chain even more complex (Hussain et al. 2018; Mathur et al. 2018).

In general, Health Care Supply Chain (HCSC) refers to the purchase of health-related products and supply is based on the demand (Mustaffa and Potter 2009; Dobrzykowski et al. 2019). It also includes ensuring the services to the patient are fulfilled on time by the hospital management (Mandal 2018). Based on the literature, the key enablers that are needed in HCSC to enable better supply chain relationships among the members, are illustrated in Fig. 9.1.

The key enablers of the health care supply chain are inventory, organisation, customer service. It is required to streamline the co-ordination between the enablers to minimise the waste and operational costs and provide timely services to the patients (Scavarda et al. 2019; Wang et al. 2020a).

- Inventory —It is necessary to optimise the inventory cost on medical wears, drugs to reduce overall health care supply chain cost.

Fig. 9.1 Enablers in Health Care Supply Chain (HCSC)

- Organisation—It is required to have a skilled workforce who can adopt new technologies and have better collaboration with both upstream and downstream supply chain members.
- Customer Services—It is a main part of the overall HCSC as the patient need to be satisfied with the proper treatment and hygienic factors.

The case organisation is one of the premier hospitals that are interested in exploring digital technologies and ready for digital transformation. Based on the discussion with the management, the roadmap to implement the digital technologies is planned in a phased manner and it is encouraging to see they are willing to invest in technology to increase the inventory visibility and satisfy the customer requirements.

The management of healthcare believes that digital technologies will reshape the HCSC in the next five years as it lays the foundation for the future strategy of the organisation. The present practices in procuring healthcare items have multiple complications in managing the supplier. Therefore, there is a need for the integration of digital technologies. Considering the emergence of new technologies and the problems in the current situation of healthcare because of COVID-19, this article assesses the influencing criteria for each enabler and proposes new digital technologies to adopt so that the overall operational efficiency of a healthcare organisation is improved. More specifically QFD based on evaluation is conducted in the case organisation to understand the importance of each criterion which affects the overall supply chain cost. Based on the outcome of results, prioritisation is done on the criteria and recommendations on the new technologies are proposed.

Based on the literature, there is a limited number of papers available on the application of digital technologies in HCSC. In addition, it is also observed that the health care organisations are not having complete trust in their supplier because of multiple flaws in the present supply chain process. Thus, the study on digital technologies in healthcare is essential to bring more value the organisations. Therefore, this study intends to contribute to the health care industry to assess the new technologies and use in the present COVID-19 condition wherever possible by addressing the following research questions.

RQ1. What are the challenges faced by HCSC and how these challenges can be addressed with technologies?

RQ2. Are digital technologies really going to help in collaborating with SC members to bring more visibility on inventory and improve their performance?

To address the mentioned research questions, the following objectives are defined

- To identify the criteria for enabler–Inventory, Organisation, Customer Service.
- To develop a framework for HCSC.
- To assess the key criteria and propose new technologies to fulfill the gap in the present system.

By answering the research objectives, this article helps organisations to know how new technologies can help in managing the HCSC process.

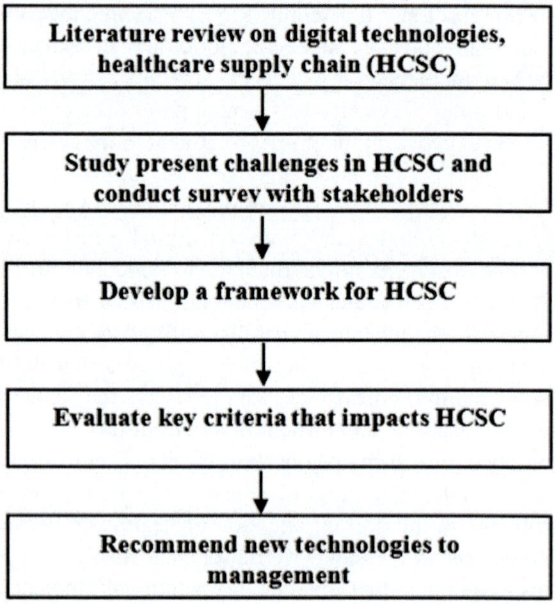

Fig. 9.2 Research methodology framework

9.2 Methodology

A literature review is conducted based on challenges faced in the healthcare supply chain. Further, the work is done on HCSC and digital technologies is analysed. Thereafter, a framework is developed for HCSC based on the literature. The QFD based approach is applied to assess the importance of each criterion in HCSC. Based on the outcome, new technologies are recommended to management. Figure 9.2 shows the flowchart of steps involved in the methodology.

9.3 Development of a Framework for Healthcare Supply Chain

The literature reveals that assessing HCSC for adopting new digital technologies was not focused on earlier and thus it is considered as a gap. Based on the literature, a new framework for HCSC is proposed as illustrated in Table 9.1. The framework is formulated from three perspectives of HCSC namely inventory, organisation, customer service to evaluate the criteria for adopting new digital technologies. The framework has two layers as tabulated in Table 9.1. The first layer contains three enablers, which influence the HCSC; the second layer comprises 16 criteria. This framework is used for assessing HCSC from digital technologies perspective.

Table 9.1 Framework for assessing HCSC for adopting new digital technologies

Enabler	Criteria	Source
Inventory	Hoarding the goods	Sarigöllü et al. (2020), Maharani and Rabbani (2020)
	Sudden demand surge for specific items	Haque et al. (2020), Polater and Demirdogen (2018)
	Item expiry	Priyan and Mala (2020), Tat et al. (2020)
	Stock-out that leads to additional procurement cost	Franco and Alfonso-Lizarazo (2020), Lucchese et al. (2020)
	Pilferage	Lawrence et al. (2020)
Organisation	Workflow design	Celesti et al. (2020)
	Staff and worker's coordination	Iyengar et al. (2020), Dobrzykowski (2019)
	Data security	Dixit et al. (2019), Ageron et al. (2020)
	Human security	Göleç and Karadeniz (2020), Khosravi and Izbirak (2019)
	Spend management	Govindan et al. (2020), Alotaibi et al. (2020)
Customer service	Hygienic and clean environment	Rizou et al. (2020)
	Transparency in treatment expense	Alotaibi et al. (2020), Buttigieg et al. (2020)
	Quick response to queries	Schoenfelder et al. (2020), Jambulingam and Kathuria (2020)
	Virtual access to frontline workers	Tan et al. (2020), Iyengar et al. (2020)
	Real-time information sharing	Ageron et al. (2020), Wan et al. (2020)
	Virtual guidance	Badri et al. (2020), Sarkis (2020)

9.4 Case Study

The study has been performed in a premier hospital situated in Tamilnadu, India (hereafter denoted as XYZ). XYZ is the best-suited organisation to conduct this study as they are already using few IT tools and matured with SCM operations. In addition, they are positive in adopting new technologies and they want to distinguish themselves as a global leader in healthcare management.

9.4.1 Assessment of HCSC Using QFD

Quality Function Deployment (QFD) is a technique being used for translation of the wants and desires of the customers into a language and specifications that can be used by organisations to develop a matrix and assess the criteria (Deepu and Ravi 2020; Wang et al. 2020b). The QFD assessment will help in subjecting any concern to Strengths, Weaknesses, Opportunities, and Threats (SWOT) analysis and remain alert and responsive to its existing/would be customers' ever demanding and changing wants and needs (Zhang et al. 2020; Mor et al. 2019). QFD approach is chosen for this evaluation as it is the best method in identifying specific sector organisation needs and helps to convert the requirements into technical specifications through the survey (He et al. 2020).

9.4.2 Construction of Criteria and Relative Weight

The evaluation of each criterion has been done based on Likert's scale with the following points.

- Completely satisfied−5,
- Relatively satisfied−4,
- Satisfied−3,
- Relatively unsatisfied−2,
- Completely unsatisfied−1.

The first column 'Importance Rate' (I_R) of the weight characteristic chart is given by the stakeholders belongs to different departments. The second column, Current level (C) shows the requirements related to the services provided at present, the stakeholders' answers to the questionnaire are displayed. The third column, Improvement Plans (I_P) displays the stakeholders' desire for the future.

The fourth column, Improvement Rate (R_I) is calculated using the relation

$$R_I = \frac{I_P}{C} \tag{9.1}$$

The weight was established according to the customer's requirements and complaints. The fifth column shows the Service point (S) where the quality characteristics representing the stakeholder's desires and needs are displayed, the Service point (S) can be described by the following statements and weights.

a. In certain service points, the stakeholder requires maximum. Attention based on certain service points that are included in the hospital expenses−1.5
b. The stakeholder states that certain attributes should be improved in the near future−1.2

c. The stakeholder considers certain attributes less important, that is, it does not matter if it changed or not−1.0.

The last two columns represent the Absolute Weight (A_W) and Relative Weight (R_w), which are calculated using the relations

$$A_W = I_R . R_I . S \tag{9.2}$$

$$R_W = \frac{A_W}{\sum A_W} x\ 100 \tag{9.3}$$

Based on the mentioned calculation, Table 9.2 is formulated and relative weight is calculated for each criterion.

9.5 Results and Discussions

The relative weight is calculated using the QFD approach and ranking is given for the criteria. The criteria identified for new technology-based improvement are illustrated in Table 9.3.

The outcome of results shows that the top six criteria have a mix of different enablers such as Inventory, Organisation and Customer Service of HCSC. The results show that the enabler 'Inventory' and 'Customer Service' are having the highest relative weight (16.54) in HCSC. So the organisation needs to focus on the corresponding criteria that are 'Sudden demand surge for specific items' and 'Virtual access to frontline workers'. This result matched with the atmosphere surfaced at XYZ.

Managerial Implications

The outcome of the new digital technology assessment has been reported to management with the improvement areas identified in HCSC and recommendations for the top six criteria are tabulated in Table 9.4. Management is reviewing the recommendations and considerations to transform into a digital organisation.

9.6 Conclusion

The COVID-19 has spotlighted the major healthcare supply chain problems on a massive scale and highlights the need for new technology deployment. It is essential to monitor the patients remotely during the pandemic situation, assess their changing needs and prioritise the treatment. In this study, a framework for the health care supply chain is developed and the criteria are analysed using QFD based approach. The framework is designed based on key enablers such as inventory, organisation, and customer service. Based on the study, 'Inventory' and 'Customer Service' are resulted as the top enabler. Both criteria 'Sudden demand surge for specific items' and 'Virtual access to frontline workers' have the highest relative weight of 16.54. Thus, these two criteria require immediate attention by the organisation to deploy the technology. The next top priority for the organisation is to focus on 'Data Security' which has a

Table 9.2 Evaluation of HCSC criteria

Sl. No	Criteria	Importance rate	Current level	Desired level	Rate of improvement	Service point	Absolute weight	Relative weight
		I_R	C	I_P	R_I	S	A_W	R_W
1.	Sudden demand surge for specific items	5	2	5	2.5	1.5	37.5	**16.54**
2.	Pilferage	4	2	3	1.5	1	6	2.64
3.	Workflow design	3	2	3	1.5	1.2	5.4	2.38
4.	Stock-out that leads to additional procurement cost	5	2	5	2.5	1.5	18.75	**8.27**
5.	Real-time information sharing	4	2	3	1.5	1.2	7.2	3.18
6.	Item expiry	3	1	3	3	1.5	13.5	**5.95**
7.	Data Security	5	1	4	4	1.5	30	**13.23**
8.	Staff and worker's co-ordination	3	1	2	2	1	6	2.64
9.	Spend Management	4	2	3	1.5	1	6	2.64
10.	Virtual access to frontline workers	5	3	5	1.67	1.5	37.5	**16.54**
11.	Workflow design	4	2	3	1.5	1	6	2.64
12.	Hoarding the goods	4	3	4	1.33	1.2	6.384	2.81
13.	Transparency in treatment expense	5	3	5	1.67	1.5	12.53	**5.52**

(continued)

Table 9.2 (continued)

Sl. No	Criteria	Importance rate	Current level	Desired level	Rate of improvement	Service point	Absolute weight	Relative weight
		I_R	C	I_P	R_I	S	A_W	R_W
14.	Human Security	4	2	3	1.5	1.2	7.2	3.18
15.	Hygienic and clean environment	5	2	4	2	1.2	12	**5.29**
16.	Quick response to queries	3	1	2	2	1.2	7.2	3.18
							Σ 226.67	

Table 9.3 Ranking based on relative weight for HCSC criteria

Rank	Enabler	Criteria	Relative weight
1	Inventory	Sudden demand surge for specific items	16.54
1	Customer service	Virtual access to frontline workers	16.54
2	Organisation	Data Security	13.23
3	Inventory	Stock-out that leads to additional procurement cost	8.27
4	Inventory	Item expiry	5.95
5	Customer service	Transparency in treatment expense	5.52

relative weight of 13.23. For the identified top three criteria, new technologies such as Big Data, the Internet of Things and Cloud Computing are proposed in HCSC.

It is pleasing to know from management that they are intended to consider the recommendations to transform their organisation into a digital healthcare supply chain organisation. With new technology upgrades, it is essential to identify the COVID-19 needs and address the concerns faster, provide more safety at an optimal supply chain cost. Besides, implementing new technologies makes the organisation be a pioneer in the healthcare industry and the demands of stakeholders are attended to greatly thereby increase in customer satisfaction.

9.6.1 Future Research Scope

This study is performed in a medium-sized health organisation and with a limited data set collected from stakeholders. The volume of data can be larger when researchers focus on bigger organisations to check if results vary with a bigger data set. Further, the current research is based on the feedback received from a case organisation in

Table 9.4 New technology recommendations for HCSC

Enabler	Criteria	Technology	Recommendations
Inventory	Sudden demand surge for specific items	Big data	– Adopt the big data-driven forecast supply chain tools to receive a demand spike alert, predict ununiform demand pattern – Explore predictive analytics to model foreseeable dependencies of the individual items
Customer service	Virtual access to frontline workers	IoT	– Evaluate technology upgrades such as IoT-based medical devices. Transform from preventive to predictive healthcare practices
Organisation	Data security	Cloud computing/IoT	– Implement robust cloud computing applications along with industrial IoT devices
Inventory	Stock-out that leads to additional procurement cost	Big data	– Use inventory management big data analytics to analyse the historical patterns to determine stockout and implement vendor-managed inventory (VMI) to enable auto-replenish the items well before stock out
Inventory	Item expiry	Cloud computing/IoT	– Use of cloud-based expiry system with IoT technology
Customer service	Transparency in treatment expense	Artificial Intelligence (AI)	– Use virtual consultant app to alert the customer by sending automatic notifications with Patients estimated expenses and actual bills with the reason for any differences

India. Researchers are encouraged to perform a similar study in any developing countries such as Argentina, Indonesia or China that are similar organisational arrangements and to validate the results post COVID-19.

References

Ageron, B., Bentahar, O., & Gunasekaran, A. (2020). Digital supply chain: challenges and future directions. *Supply Chain Forum: An International Journal, 21*(3), 133–138. Taylor & Francis.

Alotaibi, S., Mehmood, R., & Katib, I. (2020). The role of big data and twitter data analytics in healthcare supply chain management. *Smart Infrastructure and Applications*, 267–279. Springer, Cham.

Badri, S., Reddy, V. S. R. T., Chintakunta, V. R., & Kamini, S. R. (2020). A novel virtual guidance based emergency treatment using low cost drone network, in *Proceedings of the 1st ACM Workshop on Autonomous and Intelligent Mobile Systems* (pp. 1–3).

Buttigieg, S. C., Bezzina, F., Xuereb, A., & Dey, P. K. (2020). Healthcare supply chain management: Application in the maltese healthcare system. *Health Services Management Research, 33*(2), 55–65.

Celesti, A., Ruggeri, A., Fazio, M., Galletta, A., Villari, M., & Romano, A. (2020). Blockchain-based healthcare workflow for tele-medical laboratory in federated hospital IoT clouds. *Sensors, 20*(9), 2590.

Deepu, T. S., & Ravi, V. (2020). An integrated ANP–QFD approach for prioritization of customer and design requirements for digitalization in an electronic supply chain. *Benchmarking: An International Journal.*

Dixit, A., Routroy, S., & Dubey, S. K. (2019). A systematic literature review of healthcare supply chain and implications of future research. *International Journal of Pharmaceutical and Healthcare Marketing.*

Dobrzykowski, D. (2019). Understanding the downstream healthcare supply chain: Unpacking regulatory and industry characteristics. *Journal of Supply Chain Management, 55*(2), 26–46.

Franco, C., & Alfonso-Lizarazo, E. (2020). Optimization under uncertainty of the pharmaceutical supply chain in hospitals. *Computers & Chemical Engineering, 135,* 106689.

Göleç, A., & Karadeniz, G. (2020). Performance analysis of healthcare supply chain management with competency-based operation evaluation. *Computers & Industrial Engineering*, 106546.

Govindan, K., Mina, H., & Alavi, B. (2020). A decision support system for demand management in healthcare supply chains considering the epidemic outbreaks: A case study of coronavirus disease 2019 (COVID-19). *Transportation Research Part E: Logistics and Transportation Review, 138,* 101967.

Haque, M. S., Uddin, S., Sayem, S. M., & Mohib, K. M. (2020). Coronavirus disease 2019 (COVID-19) induced waste scenario: A short overview. *Journal of Environmental Chemical Engineering*, 104660.

He, L., Wu, Z., Xiang, W., Goh, M., Xu, Z., Song, W., Ming, X., & Wu, X. (2020). A novel Kano-QFD-DEMATEL approach to optimise the risk resilience solution for sustainable supply chain. *International Journal of Production Research*, 1–22.

Hussain, M., Ajmal, M. M., Gunasekaran, A., & Khan, M. (2018). Exploration of social sustainability in healthcare supply chain. *Journal of Cleaner Production, 203,* 977–989.

Iyengar, K. P., Vaishya, R., Bahl, S., & Vaish, A. (2020). Impact of the coronavirus pandemic on the supply chain in healthcare. *British Journal of Healthcare Management*, 1–4.

Jambulingam, T., & Kathuria, R. (2020). Antecedents to buyer-supplier coordination in the pharmaceutical supply chain. *International Journal of Pharmaceutical and Healthcare Marketing.*

Khosravi, F., & Izbirak, G. (2019). A stakeholder perspective of social sustainability measurement in healthcare supply chain management. *Sustainable Cities and Society, 50,* 101681.

Lawrence, J. M., Hossain, N. U. I., Jaradat, R., & Hamilton, M. (2020). Leveraging a bayesian network approach to model and analyze supplier vulnerability to severe weather risk: A case study of the US pharmaceutical supply chain following Hurricane Maria. *International Journal of Disaster Risk Reduction*, 101607.

Leite, H., Lindsay, C., & Kumar, M. (2020). COVID-19 outbreak: implications on healthcare operations. *The TQM Journal.*

Lucchese, A., Marino, A., & Ranieri, L. (2020). Minimization of the logistic costs in healthcare supply chain: A hybrid model. *Procedia Manufacturing, 42,* 76–83.

Maharani, A., & Rabbani, L. (2020). Identification of risk events and risk agents in goods hoarding at the port using fishbone diagram. *Tibuana, 3*(01), 24–29.

Mandal, S. (2018). Influence of human capital on healthcare agility and healthcare supply chain performance. *Journal of Business & Industrial Marketing.*

Mathur, B., Gupta, S., Meena, M. L., & Dangayach, G. S. (2018). Healthcare supply chain management: literature review and some issues. *Journal of Advances in Management Research.*

Mor, R. S., Bhardwaj, A., & Singh, S. (2019). Integration of SWOT-AHP approach for measuring the critical factors of dairy supply chain. *Logistics, 3*(1), 9.

Mustaffa, N. H., & Potter, A. (2009). Healthcare supply chain management in Malaysia: A case study. *Supply Chain Management: An International Journal.*

Nikolopoulos, K., Punia, S., Schäfers, A., Tsinopoulos, C., & Vasilakis, C. (2020). Forecasting and planning during a pandemic: COVID-19 growth rates, supply chain disruptions, and governmental decisions. *European Journal of Operational Research.*

Priyan, S., & Mala, P. (2020). Optimal inventory system for pharmaceutical products incorporating quality degradation with expiration date: A game theory approach. *Operations Research for Health Care,* 100245.

Polater, A., & Demirdogen, O. (2018). An investigation of healthcare supply chain management and patient responsiveness. *International Journal of Pharmaceutical and Healthcare Marketing.*

Rizou, M., Galanakis, I. M., Aldawoud, T. M., & Galanakis, C. M. (2020). Safety of foods, food supply chain and environment within the COVID-19 pandemic. *Trends in Food Science & Technology, 102,* 293–299.

Sarkis, J. (2020). Supply chain sustainability: Learning from the COVID-19 pandemic. *International Journal of Operations & Production Management.*

Sarigöllü, E., Hou, C., & Ertz, M. (2020). Sustainable product disposal: Consumer redistributing behaviors versus hoarding and throwing away. *Business Strategy and the Environment.*

Scavarda, A., Daú, G. L., Scavarda, L. F., & Korzenowski, A. L. (2019). A proposed healthcare supply chain management framework in the emerging economies with the sustainable lenses: The theory, the practice, and the policy. *Resources, Conservation and Recycling, 141,* 418–430.

Schoenfelder, J., Bretthauer, K. M., Wright, P. D., & Coe, E. (2020). Nurse scheduling with quick-response methods: Improving hospital performance, nurse workload, and patient experience. *European Journal of Operational Research, 283*(1), 390–403.

Singh, S., Kumar, R., Panchal, R., & Tiwari, M. K. (2020). Impact of COVID-19 on logistics systems and disruptions in food supply chain. *International Journal of Production Research,* 1–16.

Tan, L., Tivey, D., Kopunic, H., Babidge, W., Langley, S., & Maddern, G. (2020). Part 2: blockchain technology in health care. *ANZ Journal of Surgery.*

Tat, R., Heydari, J., & Rabbani, M. (2020). A mathematical model for pharmaceutical supply chain coordination: Reselling medicines in an alternative market. *Journal of Cleaner Production,* 121897.

Wan, P. K., Huang, L., & Holtskog, H. (2020). Blockchain-enabled information sharing within a supply chain: A systematic literature review. *IEEE Access, 8,* 49645–49656.

Wang, C. J., Ng, C. Y., & Brook, R. H. (2020a). Response to COVID-19 in Taiwan: Big data analytics, new technology, and proactive testing. *JAMA, 323*(14), 1341–1342.

Wang, H., Fang, Z., Wang, D., & Liu, S. (2020b). An integrated fuzzy QFD and grey decision-making approach for supply chain collaborative quality design of large complex products. *Computers & Industrial Engineering, 140,*.

Zhang, Y., Rysiecki, L., Gong, Y., & Shi, Q. (2020). A swot analysis of the UK EV battery supply chain. *Sustainability, 12*(23), 9807.

Chapter 10
Challenges Analysis of Leagile Supply Chains in Manufacturing Context During COVID-19

Rohit Agrawal and S. Vinodh

Abstract Developments in the field of Information Technology (IT) increases its utilisation in the manufacturing sector. Competition is increasing in the supply chain to satisfy personalised demands from customers and enhancing the relationship with the customer. To deal with problems in the conventional supply chain, industries are nowadays focusing mainly on leagile supply chain which has the capability to deal with all mentioned difficulties. The focus of this study is to analyse the challenges of leagile supply chain during COVID-19. 14 challenges have been considered from literature analysis. A Multi-Criteria Decision Making (MCDM) technique has been applied to analyse and prioritise the identified challenges. In this study, the Fuzzy VIKOR technique has been used for analysing the challenges. The conduct of the study helps the industry practitioners to critically analyse the challenges of leagile supply chain and further facilitate them to derive strategic plans to cope up with those challenges.

Keywords Leanness · Agility · Leagile supply chains · Challenges · MCDM · Fuzzy VIKOR

10.1 Introduction

Global competition and market uncertainty lead the organisation to respond quickly and efficiently. With the increasing demand and prices of resources, studies on the supply chain have become a key role to take a better decision and making effective strategies (Haq and Boddu 2017). The traditional supply chain has some difficulties in dealing with inventory cost, bullwhip effect, and quick responsiveness (Zhang et al. 2012). To deal with such problems, industries are nowadays focusing mainly on leagile supply chain which has the capability to deal with all mentioned difficulties. The concept of Leagility is significant in the current scenario. Leagility is an amalgamation of both the lean and agile paradigm. Leagile supply chain which has benefits

R. Agrawal · S. Vinodh (✉)
Department of Production Engineering, National Institute of Technology, Tiruchirappalli, Tamil Nadu, India

A. R. Sakthivel et al. (eds.), *Managing Supply Chain Risk and Disruptions: Post COVID-19*, Management and Industrial Engineering,
https://doi.org/10.1007/978-3-030-72575-4_10

of both lean principles and agile concepts has the potential to deal with a competitive environment (Haq and Boddu 2017; Soni and Kodali 2009). Leagility concept has been included in the supply chain to make effective utilisation of resources. The manufacturing industry has been impacted in several ways due to COVID-19 and the ongoing lockdown has put a lot of strain on the manufacturing industry. In the unorganised domain, the situation was much worse, as the uncertainty would impact smaller firms due to their lower profitability. This has been attributed due to lower production, lower offtake. The major proportion of employees stop coming into work, due to government directives which impacted the scale of operations and production volume. The focus of this study is to analyse the challenges of leagile supply chain in COVID-19. Several challenges have been identified such as disruption of the supply chain, increase in assembly line length because of social distancing, centralisation of facilities, high inventory level, demand uncertainty and so on (Banerjee and Mukhopadhyay 2016; Galankashi and Helmi 2016; Hamisi 2011; Manuel et al. 2015; Fadaki et al. 2019).

10.2 Literature Review

Haq and Boddu (2017) analysed the enablers of leagile supply chain for enhancing leagility of a firm. The fuzzy QFD approach was used to identify potential enablers of leagility. A case study of a food processing industry was presented and enables of leagility were analysed. This study used Fuzzy AHP and Fuzzy TOPSIS for analysing enablers of leagility. The study suggested that ‘collaborative relationship with supplier and customer’ and ‘electronic data exchange’ are the top enablers for enhancing leagility in a firm. The study helped in the implementation of leagile supply chain to achieve cost-effectiveness and responsiveness in the market.

Banerjee and Mukhopadhyay (2016) aimed to develop an approach for leagility in the supply chain. They developed a seven-stage methodological framework by considering both analytical and mathematical procedures. The study used Data Envelopment Analysis (DEA) and new design thinking approach for the identification and analysis of constraints. They presented a case study to validate the developed framework and found significant benefits on various leanness and agility parameters such as the reduction in rework and operation cost, improvement in cash flow and enhanced customer service.

Galankashi and Helmi (2016) aimed to identify and prioritise operational leagile supply chain strategies. They prioritised leagile supply chain operational activities using AHP methodology. Finally, they derived a framework showing which leagile activities are important at each stage of the supply chain.

Raj et al. (2018) aimed to analyse the supply chain performance of an organisation by considering three paradigms namely lean, agile, and leagile. They developed a conceptual framework considering leagile supply chain metrics. They used the ANP approach to compare three paradigms of the supply chain. From the result, they found that the performance of leagile supply chain is better than the lean and agile supply chain. Further, they used multi-grade fuzzy approach to evaluate the leagility index of the supply chain and to recognise weaker areas. They suggested that industry practitioners should focus on zero inventory level and new product development to deal with changing environment.

Banihashemi and Heydarnia (2018) aimed to develop an index to measure agility and leanness of any organisation and its supply chain. They developed a framework to evaluate the leanness and agility of a firm by considering various criteria from the literature study. They did a questionnaire-based survey by considering three organisations and evaluated the leanness, agility and leagile index for all three organisations. They concluded that in general, agility index was found to be better when compared with the leanness index. And, they also suggested adopting both strategies together to maintain an efficient supply chain.

Fadaki et al. (2019) aimed to analyse the drivers of the supply chain from both leanness and agility perspectives. The authors conducted a survey on 299 Australian firms by preparing a designed questionnaire. The study used the partial least square method to analyse the survey data and the result reveals that the level of uncertainty affects directly to leagility index. Further, the authors suggested implementing a balanced leagile supply chain to deal with uncertainty in demand and supply.

Li et al. (2020) aimed to analyse supplier selection by considering lean and agile perspectives. The study shortlisted the criteria for analysing suppliers from the literature study and with help of expert opinion. A total of fifteen criteria have been shortlisted for analysis. The study used the DEMATEL approach for prioritising criteria that aid in leagile supplier selection. They presented a case study of a Chinese textile manufacturing firm and their findings revealed that 'Quality' is the most important criteria for leagile supplier selection. This study helped in effective decision-making in supplier selection.

10.3 Methodology

In this study, fuzzy 'Vlsekriterijumska Optimizacija I Kompromisno Resenje' (VIKOR) is used as a solution methodology. Fuzzy VIKOR is an MCDM based method developed by Opricovic and Tzeng in the year 2004. VIKOR method focuses on prioritising alternatives by considering various conflicting criteria. VIKOR method provides compromising solutions which are close to ideal solutions. The reason for selecting Fuzzy VIKOR over other MCDM methods such as AHP, MOORA and SAW is that it identifies compromising solutions by comparing all alternatives with the ideal solution.

The steps in the Fuzzy VIKOR method are as follows (Shemshadi et al. 2011):

Step 1: Define the objective of the study and identify the challenges which need to be analysed

Step 2: Select suitable criteria to analyse the identified challenges

Table 10.1 The fuzzy scale used in the study

Scale for criteria weights		Scale for challenges rating	
Linguistics	Fuzzy number	Linguistics	Fuzzy number
Very poor	(0,0,0.1)	Very low	(0,0,1)
Poor	(0,0.1,0.3)	Low	(0,1,3)
Fairly poor	(0.1, 0.3, 0.5)	Medium low	(1,3,5)
Fair	(0.3, 0.5, 0.7)	Medium	(3,5,7)
Fairly high	(0.5, 0.7, 0.9)	Medium Good	(5,7,9)
High	(0.7, 0.9, 1)	Good	(7,9,10)
Very high	(0.9, 1, 1)	Very good	(9,10,10)

Step 3: Formulation or adoption of a Fuzzy scale to collect data from experts.
The scale used in this study is mentioned in Table 10.1.

Step 4: Defuzzification of criteria weight and challenges rating.
In this study, Eq. 10.1 has been used for the defuzzification of collected data.
For a TFN $x_{ij} = \left(l_{ij}, m_{ij}, u_{ij}\right)$.
Defuzzification of a fuzzy number can be calculated as (Agrawal et al. 2017):

$$y_{ij} = \frac{l_{ij} + (4 * m_{ij}) + u_{ij}}{6} \tag{10.1}$$

Step 5: Calculation of aggregate criteria weights and challenges rating.

In this step, data from different experts could be combined to get aggregate criteria weights and challenges rating using Eq. 10.2.

$$z_{ij} = \frac{\sum_{j=1}^{n} y_{ij}}{n} \tag{10.2}$$

Where, n is number of experts.

Step 6: Calculation of best and worst values of challenges. It is represented by

$$f_j^* = \begin{pmatrix} \max\left(z_{ij}\right) & for\ benefit\ criteria \\ \min\left(z_{ij}\right) & for\ cost\ criteria \end{pmatrix} \tag{10.3}$$

$$f_j^- = \begin{pmatrix} \min\left(z_{ij}\right) & for\ benefit\ criteria \\ \max\left(z_{ij}\right) & for\ cost\ criteria \end{pmatrix} \tag{10.4}$$

Step 7: Calculate utility value (S_i), regret value (R_i) and VIKOR index value (Q_i)

$$d_{ij} = R_i = \frac{\left(f_j^* - z_{ij}\right)}{\left(f_j^* - f_j^-\right)} \tag{10.5}$$

$$S_i = \sum w_j * d_{ij} \tag{10.6}$$

$$R_i = \max\left(w_j, d_{ij}\right) \tag{10.7}$$

$$Q_i = \frac{v_j(S_i - S^*)}{(S^- - S^*)} + \frac{(1 - v_j)(R_i - R^*)}{(R^- - R^*)} \tag{10.8}$$

Where,
$S^* = \min(S_i)$
$S^- = \max(S_i)$
$R^* = \min(R_i)$
$R^- = \max(R_i)$
v_j is maximum strategy group weight
$(1 - v_j)$ is weight of individual regret

Step 7: Ranking of challenges are based on Q values. The challenges which got minimum Q value is ranked first.

Suggesting compromising solution:

Challenges are ranked based on increasing order of Q value as shown in step 7 only if it follows two conditions.

Condition 1: Acceptable advantage: if $Q\left(C^{(2)}\right) - Q\left(C^{(1)}\right) \geq \frac{1}{K-1}$ where $C^{(1)}$, $C^{(2)}$ shows first and second position of challenges based on Q value and K denotes number of challenges.

Condition 2: Acceptable stability: the challenge $Q\left(C^{(1)}\right)$ is stable in decision making, if it got the same position based on R and S values also.

If any of the conditions are not satisfied, then there exists a compromising solution.

1. Challenges $C^{(1)}$ *and* $C^{(2)}$ will act as a compromising solution, if only condition 2 is not satisfied.
2. Challenges $C^{(1)}$, $C^{(2)}$, $C^{(3)}, \ldots,$ $C^{(K)}$ will act as a compromising solution if only condition 1 is not fulfilled.

Where, $C^{(K)}$ is found by $Q\left(C^{(K)}\right) - Q(C^{(1)}) < \frac{1}{(K-1)}$ for maximum 'K'.

10.4 Case Study

In this study, challenges of leagile supply chain have been analysed. From the literature review, fourteen potential challenges have been identified pertaining to leagile supply chain strategies. The list of considered challenges is presented in Table 10.2.

After identifying challenges of leagile supply chain, this study identifies several criteria to analyse challenges. In this regard, this study considers five crucial criteria

Table 10.2 Identified leagile supply chain challenges

S.No.	Challenges	References
1	Failure to meet the scheduled shipping date (C_1)	Hamisi (2011), Manuel et al. (2015), Banerjee and Mukhopadhyay (2016), Galankashi and Helmi (2016), Fadaki et al. (2020)
2	Long lead times for order confirmation from production (C_2)	
3	Change orders after order booking leading to disruption in manufacturing (C_3)	
4	Culture of innovation and improvement in the company (C_4)	
5	Inability to consolidate and also rework during shipment (C_5)	
6	Inability to understand and meet customer requirements (C_6)	
7	Rework/effort wastage/Longer lead time (C_7)	
8	Cope with the demand variation (C_8)	
9	Implementation of Electronic Data Interchange (C_9)	
10	Complexities associated with global sourcing (C_{10})	
11	High inventory levels (C_{11})	
12	Outdated and/or non-integrated technologies (C_{12})	
13	Disruption of the supply chain (C_{13})	
14	Increase in assembly line length because of social distancing (C_{14})	

which are important in leagile supply chain practices. The criteria are price, quality, time, flexibility, and ability to collaborate.

After finalising criteria, input sheets have been prepared to collect data for analysing leagile supply chain challenges. A team of four experts has been formed and input sheets have been sent to them for data collection.

The collected data is defuzzified using Eq. 10.1 and aggregate rating of challenges has been calculated. To calculate the aggregate rating of challenges and weights of criteria, Eq. 10.2 is used.

The aggregate rating of challenges and weights of criteria is presented in Table 10.3.

The next step is to identify the best and worst values of each criterion. The most important challenge for a particular criterion is termed as its best challenge value. Similarly, the least important challenge for a particular criterion is termed as its least challenge value. For each criterion, the best value and worst value have been identified using Eqs. 10.3 and 10.4. The identified best value and worst value for each criterion are presented in Table 10.4.

Table 10.3 Identified leagile supply chain challenges

Criteria	Price	Quality	Time	Flexibility	Ability to collaborate
Weights	0.862	0.837	0.6	0.792	0.696
Challenges rating	Price	Quality	Time	Flexibility	Ability to collaborate
C_1	8.625	6.458	7.417	7.417	7.917
C_2	7.917	6	8.375	7.917	6
C_3	9.083	5.5	9.333	6.917	6.958
C_4	6.958	9.333	8.625	8.167	6.958
C_5	9.083	6.958	6.917	7.917	6
C_6	6.958	6.958	7.917	8.625	7.917
C_7	7.917	6.917	7.917	6.958	7.917
C_8	6.958	5	6.958	6	6.958
C_9	7.917	7.917	7.917	7.417	7.417
C_{10}	8.625	6.917	7.917	7.917	6.958
C_{11}	7.917	6.917	6.458	7.417	6
C_{12}	7.917	6.958	6.917	6.958	6.958
C_{13}	7.417	5.958	6.958	6.958	6
C_{14}	9.333	7.917	6.458	5	6.958

Table 10.4 Identified best and worst value of each criterion

Criteria	Price	Quality	Time	Flexibility	Ability to collaborate
f^*	6.958	9.333	9.333	8.625	7.917
f^-	9.333	5	6.458	5	6

After identifying the best and worst values of all challenges, the next step is to calculate utility value (S_i), regret value (R_i). Equations 10.5–10.7 have been used to calculate utility value and regret value. Table 10.5 presents utility value and regret value for each challenge.

Further VIKOR index value (Q_i) to be found. For calculating VIKOR index value (Q_i), Eq. 10.8 has been used. To calculate the VIKOR index, the first values of S^*, S^-, R^*, and R^- have been calculated as shown below:

$S^* = \min(S_i) = 0.596$

$S^- = \max(S_i) = 2.875$

$R^* = \min(R_i) = 0.348$

$R^- = \max(R_i) = 0.862$

Then value of v_j, that is, strategy group weight is specified as 0.5 based on a study from previous literature on leagile supply chain. Then finally VIKOR index is calculated based on Eq. 10.8. Ranking of challenges have been done based on Utility value (S_i), Regret value (R_i), and VIKOR index as depicted in Table 10.6.

Table 10.5 Identified leagile supply chain challenges

Challenges	Utility value (S_i)	Regret value (R_i)
C_1	1.824	0.605
C_2	2.042	0.696
C_3	2.233	0.771
C_4	0.596	0.348
C_5	2.585	0.771
C_6	0.755	0.459
C_7	1.474	0.467
C_8	2.254	0.837
C_9	1.363	0.348
C_{10}	1.87	0.605
C_{11}	2.374	0.696
C_{12}	2.023	0.504
C_{13}	2.374	0.696
C_{14}	2.875	0.862

Table 10.6 Ranking of challenges

Challenges	Utility value (S_i)	Rank	Regret value (R_i)	Rank	VIKOR index (Q_i)	Rank
C_1	1.824	5	0.605	7	0.52	6
C_2	2.042	8	0.696	10	0.655	8
C_3	2.233	9	0.771	12	0.771	11
C_4	0.596	1	0.348	1	0	1
C_5	2.585	13	0.771	12	0.848	13
C_6	0.755	2	0.459	3	0.143	2
C_7	1.474	4	0.467	4	0.308	4
C_8	2.254	10	0.837	13	0.839	12
C_9	1.363	3	0.348	2	0.168	3
C_{10}	1.87	6	0.605	7	0.53	7
C_{11}	2.374	12	0.696	10	0.728	10
C_{12}	2.023	7	0.504	5	0.465	5
C_{13}	2.374	11	0.696	10	0.728	9
C_{14}	2.875	14	0.862	14	1	14

The final step is the derivation of a compromising solution. Challenges are ranked based on increasing order of Q value as shown in step 7 only if it follows two conditions as mentioned in methodology.

Condition 1: $Q\left(C^{(2)}\right) - Q\left(C^{(1)}\right) \geq \frac{1}{K-1}$; $0.143 - 0 \geq \frac{1}{14-1}$; $0.143 \geq 0.077$; hence satisfied condition 1.

Condition 2: the challenge $Q(C^{(1)})$ is stable in decision making if it got the same position based on R and S values also. From Table 10.6, it can be found that the fourth challenge occupied the top position from all perspectives including R, S and Q values. Hence, it concludes condition 2 also gets satisfied.

As both conditions are satisfied, the final ranking of challenges is based on the VIKOR index (Q_i) value.

10.5 Results

In this study, challenges of leagile supply chain have been analysed using the fuzzy VIKOR method. From the analysis, it is found that 'Culture of innovation and improvement in the company (C_4)' is the most vital challenge for leagile supply chain. 'Inability to understand and meet customer requirements (C_6)' and 'Implementation of Electronic Data Interchange (C_9)' are the second and third important challenges for leagile supply chain.

The priority order of the challenges is $C_4 > C_6 > C_9 > C_7 > C_{12} > C_1 > C_{10} > C_2 > C_{13} > C_{11} > C_3 > C_8 > C_5 > C_{14}$ as presented in Table 10.7.

The identification of topmost leagile supply chain challenges helps industry practitioners to make effective strategic plans to overcome those challenges.

Table 10.7 Final ranking of challenges

Challenges	VIKOR index (Q_i)	Ranking
C_1	0.52	6
C_2	0.655	8
C_3	0.771	11
C_4	0	1
C_5	0.848	13
C_6	0.143	2
C_7	0.308	4
C_8	0.839	12
C_9	0.168	3
C_{10}	0.53	7
C_{11}	0.728	10
C_{12}	0.465	5
C_{13}	0.728	9
C_{14}	1	14

10.6 Conclusions

COVID-19 pandemic has affected the almost entire world and because of this, manufacturing sectors are suffering a lot. It has heavily affected the supply chain because of transportation issues, labour issues arise due to COVID-19. Hence, the current scenario required advanced principles to deal with all such issues. In order to facilitate manufacturing firms to gradually stabilise production performance and ultimately to attain business excellence, Leagile supply chains with enhanced focus could be a promising solution. In this viewpoint, the study focused on the analysis of challenges of leagile supply chains using the MCDM approach during COVID-19. The priority order of challenges is being derived. Based on the study, it is found that 'Culture of innovation and improvement in the company (C_4)', 'Inability to understand and meet customer requirements (C_6)' and 'Implementation of Electronic Data Interchange (C_9)' are top challenges for leagile supply chain.

References

Agrawal, R., Asokan, P., & Vinodh, S. (2017). Benchmarking fuzzy logic and ANFIS approaches for leanness evaluation in an Indian SME. *Benchmarking: An International Journal, 24*(4), 973–993.

Banerjee, A., & Mukhopadhyay, S. K. (2016). A contemporary TOC innovative thinking process in the backdrop of leagile supply chain. *Journal of Enterprise Information Management, 29*(3), 400–431.

Banihashemi, S. A., & Heydarnia, Z. (2018). A Hybrid Model for Evaluating Leagile Supply Chain Performance in Industry. *Journal of Industrial Strategic Management, 3*(1), 43–54.

Fadaki, M., Rahman, S., & Chan, C. (2019). Leagile supply chain: design drivers and business performance implications. *International Journal of Production Research, 58*(18), 5601–5623.

Fadaki, M., Rahman, S., & Chan, C. (2020). Leagile supply chain: design drivers and business performance implications. *International Journal of Production Research, 58*(18), 5601–5623.

Galankashi, M. R., & Helmi, S. A. (2016). Assessment of hybrid Lean-Agile (Leagile) supply chain strategies. *Journal of Manufacturing Technology Management, 27*(4), 470–482.

Hamisi, S. (2011). Challenges and opportunities of Tanzanian SMEs in adapting supply chain management. *African Journal of Business Management, 5*(4), 1266.

Haq, A. N., & Boddu, V. (2017). Analysis of enablers for the implementation of leagile supply chain management using an integrated fuzzy QFD approach. *Journal of Intelligent Manufacturing, 28*(1), 1–12.

Li, Y., Diabat, A., & Lu, C. C. (2020). Leagile supplier selection in Chinese textile industries: a DEMATEL approach. *Annals of Operations Research, 287*(1), 303–322.

Manuel, P., Al-Hamadi, H., & Qureshi, K. (2015). Challenges, strategies and metrics for supply-driven enterprises. *Annals of Operations Research, 233*(1), 293–303.

Raj, S. A., Jayakrishna, K., & Vimal, K. E. K. (2018). Modelling the metrics of leagile supply chain and leagility evaluation. *International Journal of Agile Systems and Management, 11*(2), 179–202.

Shemshadi, A., Shirazi, H., Toreihi, M., & Tarokh, M. J. (2011). A fuzzy VIKOR method for supplier selection based on entropy measure for objective weighting. *Expert Systems with Applications, 38*(10), 12160–12167.

Soni, G., & Kodali, R. (2009). Performance value analysis for the justification of the leagile supply chain. *International Journal of Business Performance Management, 11*(1–2), 96–133.

Zhang, Y., Wang, Y., & Wu, L. (2012). Research on demand-driven leagile supply chain operation model: a simulation based on anylogic in system engineering. *Systems Engineering Procedia, 3,* 249–258. https://doi.org/10.1016/j.sepro.2011.11.027.

Chapter 11
Impact of COVID-19 on Global Supply Chains and the Role of Digitalisation: A VUCA Approach

Kotha Raj Kumar Reddy and P. Kalpana

Abstract The alarming spread of the COVID-19 pandemic has made the entire world anxious. It has not only created a global health crisis but has also resulted in a poorly performing global economy. The entire world is investing their research and development team into inventing a vaccine and the extensive process of vaccine development seems to be taking a lifetime. Nevertheless, these tough times have transformed the functioning of business activities and firms including work from home across the world, operation with minimal staff in a firm's location, etc. Therefore, this chapter focuses on demonstrating the scenario of post-COVID supply chains. In this context, we must focus on sustainable and efficient supply chains to lead the business from the front. Here, our aim is to discuss the challenges in supply chains in the post-COVID era, the role of digital technologies, the essence of risk assessment, and industry 4.0 supply chains. Digital technologies such as analytics, blockchain, artificial intelligence, big data, cloud computing, and quantum computing hold the key to ensure future supply chain security, safety, sustainability, visibility, and transparency. Also, we have discussed VUCAness and its essence, and risk assessment frameworks.

Keywords Global supply chain · Digitalisation · Risk assessment · VUCA approach

K. R. K. Reddy
Supply Chain Analytics - SME, Chennai, Tamil Nadu, India

P. Kalpana (✉)
Department of Mechanical Engineering, Indian Institute of Information Technology Design & Manufacturing (IIITDM), Kancheepuram, Tamil Nadu, India
e-mail: kalpana@iiitdm.ac.in

A. R. Sakthivel et al. (eds.), *Managing Supply Chain Risk and Disruptions: Post COVID-19*, Management and Industrial Engineering,
https://doi.org/10.1007/978-3-030-72575-4_11

11.1 Introduction

Currently, the entire world is experiencing a difficult time due to the COVID-19 (SARS - CoV-2) pandemic. As per the records, COVID-19 is said to have originated in Wuhan and within no time it has emerged as a global pandemic. WHO (World Health Organization) declared COVID-19, a global pandemic on March 11, 2020, and efforts all over the world are being taken to understand this novel virus. As per 11/08/2020 confirmed COVID cases crossed 20 million cases (Global COVID-19 information 2020) with over 7 lakh deaths across the world. Although a large population of patients is recovering (~13 million), it is still a deadly virus that has forced various governments across the nations to enforce partial lockdown/safety restriction in place to contain the spread of this fast pandemic due to human transmission. As a result, few restrictions on the imports and exports have been enforced until the situation gets better. As almost the entire world fights hard with COVID-19, the officials are focusing on mitigating the spread of the virus by tracing infected people efficiently as soon as possible and isolating them by maintaining a quarantined environment to combat this deadly virus, which helps a lot to reduce the potential hotspots. Epidemiologists, Virologists, and Immunologists are keenly studying the existing mathematical models and techniques to deal with COVID better. In this critical time only, few are thinking out of the box and focusing on building a system that can absorb these disruptions. COVID-19 has become a part of this society and some say it will stay for a long time. As a result, it seems tough to get back to the normal pre-COVID days and may require us to define a new normal, and that normal will be varying time to time. There are many challenges for the supply chain due to this pandemic, that is material shortage, delivery delays, performance degradation in terms of revenue, decrease in service level and productivity, and disruptions in supply, demand, inventory, and logistics. Global supply chains have been impacted severely due to this pandemic. The outbreak in Wuhan resulted in almost 51000 companies across the world which had direct suppliers in Wuhan getting heavily impacted. Due to this pandemic, supply chains are experiencing unprecedented disruptions in supply, demand, and transportation and it is impossible to solve the critical challenges by generalising supply chains, because each supply chain is unique. For example, the automotive industry may get a boost again because of mandating social distancing and it will be difficult for farmers to sell their crops during this pandemic and post-pandemic too. In supply chains, there are a few ambiguities, that is how long the SC can sustain a disruption, how long it takes for an SC to recover from the pandemic, standard operating policies will complicate the operation and those require certain changes for smooth transactions.

Here we are trying to study the impact of COVID-19 on various supply chains, and also try to get an idea of how supply chains are going to behave after this COVID-19 pandemic, what will be key challenges for supply chains post-COVID-19. Key research questions we are trying to answer are as follows:

1. How to deal with supply chain VUCA?

2. Which industries have been impacted most and least by the COVID-19 pandemic?
3. What will be the new challenges for SC due to the on-going pandemic?
4. What are the long-term impacts on SC's due to the COVID-19 pandemic?
5. What are the key areas to improve, to achieve maximum efficiency in a new normal environment for supply chains?
6. What SC practitioners can do in order to ensure smooth and frictionless operations?
7. Is digitalisation and adopting emerging technologies going to benefit SC activities?
8. How to tackle similar disruptions soon?
9. What strategies can lead to improved efficiency of SC activities post-COVID-19?

Companies should understand the necessary changes to be made or actions to be taken to mitigate the risk against the disruptions. The on-going pandemic will create new problems which need organisational attention, as well as creating a new scope for business, if an organisation (can be well established or medium) should realize the need, demand, opportunities, and disruptions or else they will be outplayed by the competitors and new entrants in the industry. Organisations should go back and revise the strategy, vision, and mission statements again. COVID-19 pandemic has tested the various systems and business processes, if the product or service can be maintained affordably with the adoption of a novel method for their business processes for a smooth and frictionless environment for the future, even in tough times like COVID-19.

11.2 Risk Assessment (Black Swan & Perfect Storms)

Two terms in risk management that got researchers'/practitioners' attention these days are "black swans" and "perfect storms" which help to describe the unthinkable or the extremely unlikely which have never been encountered before. The universal truth is that the swans are known to be white in colour, and there is very little chance of them being black in colour, that is black swans can be treated as outliers. So, a black swan is an event that is even impossible to imagine, because we do not normally encounter and is highly unlikely (Paté-Cornell 2012). For example, the on-going COVID-19 crisis is the epitome of the black swan. Perfect storms are also like black swan disasters, but there will be a chance to predict the disaster in advance, considering the previous experience or current information. For example, the 9/11 attacks were not black swans, they were perfect storms since there was a scope of predicting the attack as the FBI have known to have some clues in advance regarding the attacks (Black swan and perfect stroms 2012).

Risk assessment is trivial in terms of systems, their functional components, interdependencies, and standard operating procedures (SoP) for any firm to have a better understanding of their stand. The risk assessment is essential to redesign the existing processes/procedures to enhance flexibility and adaptability. Let's discuss two examples to understand the importance of risk assessment.

1. In recent times sanitisation has become very essential to ensure health safety. As a result, many apartments and firms have started ensuring proper sanitisation norms. People also have been carrying sanitizers with them and there have been cases of vehicles catching fire. Similarly, the supporting staff sanitises COVID isolation centres and there have been reported cases of electrical fire accidents due to callous use of sanitizers. The main problem is in the ignorance of people of the new processes and the associated risks while adapting or implementing a new system or process, therefore leading to hazardous accidents.
2. Poor governance, especially neglecting organisational indicators, process indicators, and behavioural indicators, is another factor. Many disasters/failures have occurred in firms due to a lack of attention to the mentioned indicators.

During such crises, it is imperative for any organisation to validate and improve its agility and visibility in their SC's, regardless of the industry type and it becomes critical to leverage data-driven decision making for risk mitigation & resilience in SC. The aim is to improve the visibility, transparency, resilience, and agility of SC.

11.2.1 VUCA

Across industries, the rise of VUCA (volatility, uncertainty, complexity, and ambiguity) is leading to change in the market and competition. VUCA is an acronym that directly challenges the way leaders develop stabilities and direction to businesses, VUCA stands for volatility, uncertainty, complexity, and ambiguity, and is used to describe an environment and is also a trendy way of depicting unpredictable change (Bennett and Lemoine 2014). Though the four words will have different meanings, in a business context these are synonyms, after all, "How do you plan anything in such a VUCA world?" matters.

Supply chains are no exception for the VUCA world; supply chains are complex networks that involve multiple organisational presences with support/follow-up from OEM internal employees. In most of the SC's, there is insufficient data about processes and products to comprehensively understand the situation while in some cases; the understanding acquired by the existing data might not be of any value in predicting instances or failures. It is thus required to have a good combination of information and experimentation to understand the existing methods and infrastructure. So, the increase in complexity will result in increased volatility, uncertainty, and ambiguity or vice versa.

1. **Volatility**—It provides a profit opportunity and it is a new reality, volatility implies a relatively unstable change, that is data is available, and the situation is understandable, but change is frequent and still unpredictable even after having

enough data and an understandable situation. Agility will be a key to tackle volatility, resources and assets should be aggressively directed for potential future flexibility. The equipment/assets are not meant to be sitting ideal, the working environment should communicate details with each node.

a. Example: Price volatility is a part of managing supply chains, it is very unpredictable, and it can be a direct risk for supply chain inefficiency. Through agility in supply chain activities, supply chains can mitigate volatility to a certain extent.

2. **Uncertainty**—It is an opportunity, uncertainty defines a lack of knowledge as to whether an event results in unwelcome consequences, cause and effects are studied and understood, but it is still unclear whether an event will create a significant change. Information is critical to reducing uncertainties in the environment and one should focus on increasing information sources and validate the authenticity of the information. Organisations should focus on digitised industry 4.0 supply chains which can yield a lot of information about the processes and transactions.

   a. Example: Uncertainty is always a part of firm's decision making but this has become worse with the pandemic. Supply chain uncertainty is mostly associated with decision-making in unprecedented events. The transparency and visibility of information are key to reduce uncertainty in supply chain activities post COVID. Potential problems to look for post-COVID are
      i. What and how I'm going to deliver to customers?
      ii. Is there any novel way for my current business processes, that is, the pre-COVID business model to adapt to a post-COVID setup too?
      iii. Price affordability and customer retaining, if we are adopting any novel method for business processes or services.
      iv. What will be the quality problems that the customers might face in the post-COVID scenario?
      v. How to keep decision makers informed about the activities post-COVID for better transparency and visibility?
      vi. New product launches and dealing with product launches of competitors.

3. **Complexity**—Simplifying complexity is a major opportunity to make a difference in the industry, complexity is defined by an elaborate interconnected network of information and procedures in the environment. Restructuring internal operations & responsibilities to match external complexity is an efficient way to address this complexity. Organisations must attempt to create a tight synchronisation between internal capabilities and external requirements.

   a. Example: Supply chain complexity is caused by interdependency, uncertainty, variability, variety, etc. As discussed earlier one cause is the generation of others. Key complexities to look for post COVID are

   i. Tailored and stringent delivery expectations of customers and suppliers
   ii. Faster lead times, safety, and trust in manufacturing
   iii. Expanded products and services
   iv. Expanding markets to new locations
   v. New acquisitions and identifying potential new entrants for acquisition.

4. **Ambiguity**—Ambiguity equals opportunity, ambiguity arises when cause and effect are not clearly understood and there is no precedence defined across events for making predictions. Experimentation is a necessary requirement for mitigating ambiguity in an environment. Only through intelligent experimentation can explain what strategies are beneficial and what is going wrong in firm's tough times, through these firms can update rules of business.

   a. Example: Lack of clarity of situation in instances of the supply chain leads to ambiguity. It can be the unclear regulatory environment, disruptions in the industry, the preparedness of firms to absorb disruptions, etc.

11.3 Potential Challenges for Supply Chains Post-COVID-19

1. **Cyber security**: With the increased digitisation and remote working culture there is potential risk of data theft and fraudulent activities. So, firms must invest in cybersecurity in order to secure their information.
2. **Systems integration, intra and intercompany**: For efficient decision making and to reduce uncertainties the intra and intercompany systems must be integrated.
3. **Environment projection and safety regulations**: After a pandemic, environmental and safety precautions must be followed in order to maintain the sustainability and safety of stakeholders.
4. **Job skills and knowledge transfer**: After these tough times companies will realize their leaking instances, so new job profiles will be required in a firm in a bulk.
5. **International trade requirements and regulatory changes**: Shift from globalisation to regularisation
6. **Autonomous vehicles**: This was a far-hanging fruit for several years, now it is becoming reality. In these tough times disruption in the automotive industry will be a big blow for many traditional suppliers, and adapting is even tough for them.
7. **Transport modality shifts**

How Will the Post-COVID-19 Situation Be?

Supply chains are becoming extremely complex and global in reach. However, most of the firms are unaware of their deep supply chain network; most of the firms have control till their tier-1 suppliers only. The pandemic has made firms to realize that most of their suppliers depend upon China for raw materials and spare parts. For example, the automotive industry has 80% of its supply chains to China in some way or the other.

Here, let us discuss few supply chains for a better understanding.

11.3.1 Automotive Supply Chain

Automotive supply chains are complex networks, consisting of a large number of components with a giant infrastructure and various stakeholders including suppliers, dealers, governing bodies, manufacturers, employees, and customers. Every step in the automotive supply chain is very much sensitive because of various stakeholders' presence, large-scale investment, and infrastructure (Biswas and Gupta 2019). Automotive supply chains are well defined and well-controlled through proved practices along with IoT introduction. Because of the pandemic, car sales have reduced a lot during the lockdown, and it is expected to rise in near future because of mandated social distancing norms. So, the market for automotive products will be there, and the challenge lies in managing efficient operations and supply chains.

Challenges for Automotive Supply Chains Post-COVID:

1. Stockholders are not well informed about the inventory, demand, and transaction details.
2. Since stakeholders are not well-informed inventory obsolesces are frequent
3. Lack of informed decision making (Bai and Sarkis 2020)
4. Existing systems are not robust to data duplications, noisy data produces a lot of inconsistent alert system and the same data is being used for mining insights (Attaran and Gunasekaran 2019)
5. Transparency & Traceability is not maintained throughout the supply chain (Bürer et al. 2019)
6. Limited data availability, which eventually leads to the bullwhip effect, especially in uncertain times like COVID, the effect will be more (Dolgui et al. 2020)
7. Tough to manage the large supplier base and complex chains in vulnerable times with imposed safety norms and restrictions (Fu and Zhu 2019).

OEM's Should Focus On:

1. Deep supply chain awareness
2. Early detection capability of vulnerabilities

3. Risk modelling capability, efficient leverage of proven frameworks and acronyms
4. Moving digital and adopting digital technologies.

11.3.2 Agriculture Supply Chain

Agriculture supply chains are more like fast-moving consumer goods (FMCG) supply chains in many ways but differ in terms of raw material and the final product (Sharma et al. 2020). The raw materials are procured from fields by paying the equivalent money to customers, and the product is processed and packed for human or animal consumption. Agriculture supply chains include operations such as pre-production, production, storage, processing, retail, and distribution before the final product reaches the customer. Because of the pandemic, agriculture supply chains were affected a lot because of enforcing lockdown and safety precautions. The materials were not procured well during these tough times, since lack of transportation and worsening demand. In the agriculture supply chain again, there are two divisions', that is crops for commercial usage and for daily consumption. Pandemic created a lot of uncertainty around agriculture supply chains, commercial needs are being compromised due to unavailability of stocks in nearest stores, and trust and safety restrictions became barriers for vegetables and fruits.

Challenges for Agriculture Supply Chains Post-COVID:

1. Lack of sustainability in agriculture supply chains
2. Farmers are not aware of the downstream member's needs and market behaviour, which creates a slack.
3. Lack of industrialisation in agriculture supply chains
4. Information asymmetry
5. Poor product traceability
6. Every node in the agriculture supply chain is not seamlessly connected or not well informed about other nodes; this increases uncertainty in decision making.

Agriculture Supply Chain Regulators Should Focus On:

1. Understanding complex agriculture ecosystems
2. Including farmers into the ecosystem to keep them informed about the demand and vulnerabilities
3. Focus on forecasting and new market methods
4. Introducing new market and delivery methods.

11.3.3 Healthcare Supply Chain (Cold Chains)

Cold chain denotes the series of actions and equipment used to maintain the products within the specified temperature range from production/harvest to consumption and it is also known as controlled temperature supply chain. The cold chain is used to preserve and extend the shelf life of products such as confectionery, vegetables, meat & marine products, fruits, and dairy products. Vaccine cold chain is associated with maintenance of temperature of vaccines from point of manufacture to the end customer. The reverse cold chain is a controlled temperature chain in the reverse direction, that is from downstream members to upstream members, for example blood samples collected from patients or people must be maintained in the cold chain for effective usage. Vaccine cold chains must ensure that people have access to the right vaccine in good condition, at the right time, in the right place. Currently, many institutes are performing a trial for their vaccines; if a vaccine clears all stages of testing then it will be approved for treatment. It is tough to ensure the vaccine reaches each corner of the world. If vaccines are not in specified temperature for more than allowable time, vaccines must undergo quality checking to ensure the right condition, which is a tedious process.

Challenges for Cold Chains:

1. Manufacturability and mass manufacturing of vaccines within no time to meet the demand.
2. Insufficient forecasting systems for specific diseases
3. Risk of reduced potency of vaccines administered, poor availability of immunisation supplies (Improving cold chain systems 2020)
4. Lack of effective system for monitoring temperature in logistics and logistics are not planned in an optimal way (Immunization supply chain and logistics 2020)
5. Not able to accommodate sudden vaccine demand changes
6. Lots of uncertainties in distributions of vaccines, when compared to the planned distribution
7. Lack of end to end cold chain visibility in real-time
8. Lack of transparency and traceability in cold chain activities
9. Stockholders are notified about instances after their occurrence, not in near real-time. Performing exploratory data analysis (EDA) on collected data after the process will be equal to post-mortem analysis (Global COVID-19 information 2020)
10. Inadequate implementation monitoring systems and management (Immunization supply chain and logistics 2020)
11. Ineffective vaccine disposal system after use or after its life
12. No practice of knowledge-based vaccine distributed system to downstream members (Who is preparing the next generation of immunization supply chain professionals 2020).

Regulators Should Focus On:

1. Digitising the temperature monitoring equipment and the logistics
2. Leveraging digital technologies to improve visibility, transparency, and efficiency (Handbook on eVIN application for vaccine and cold chain manager India (2016)).

3. Reducing the uncertainty by collecting a lot of data about vaccines in every stage and process.
4. Focusing on reverse logistics
5. Ensuring last-mile vaccine safety, since health workers carry vaccines in portable cold chambers ensuring vaccine potential during the last mile is essential.

11.3.4 Industry 4.0 Supply Chains

Supply chains need to be agile, after this pandemic supply chain regulators should focus on the following:

1. ***Greater Customisation Through Additive Manufacturing***
    - Additive manufacturing or 3D printing is a booming field now, and it is affecting traditional supply chains significantly. As many know it is the process of making products by deposing material layer by layer. Though it is trendy and not seen its full potential, it must reach its maximum potential soon.
    - After this pandemic cost-cutting will be a key parameter for any organisation to sustain in the respective industry. So, prototyping models with additive manufacturing will cost less than traditional techniques, whereas mass manufacturing with 3D printing is not cost-effective.
    - Additive manufacturing models can be done on as it is having shorter cycle times.
2. ***End-to-End Integration of Advanced Analytics***
    - Through IoT and other data acquisition techniques a lot of data is being captured about raw material, manufacturing processes, finished product, logistics, customer usage patterns, etc.
    - Data is the new oil and significantly impacts business processes, data won't add any value to the business processes.
    - Post-COVID, industry 4.0 supply chains are essential, it uses advanced techniques such as high-performance computing, predictive & prescriptive analytics, and big data to mine for insights and information to stakeholders in near real-time, which enhances the preparedness of the supply chain.
3. ***Visibility and Transparency for all Stakeholders***
    - Now, every component of the supply chain is not interconnected, and can't communicate information across nodes efficiently, though third parties are there to coordinate supply chain activities, those data is also properly maintained, there exists a golden opportunity of creating a unique point of truth, which enhances the informed decision making.

- Unique point of truth also enhances supply chain visibility, transparency and traceability from the raw material to the end user.

4. ***A Move Beyond Postmodern ERP***

- The conventional Enterprise resource planning (ERP) systems are rigid in nature, and tough to achieve near real-time communication. So, organisations must focus on building advanced IT capabilities to enhance their business operations. Robustness, integrability, connectivity, service index, agility, and maintenance are key for IT infrastructure.

5. ***Widespread Incorporation of Digitisation of Equipment***

- Now most of the manufacturing facilities, logistics and other activities are digitised using technologies like IoT, IIoT, and other data acquisition techniques. Post COVID-19, the level of digitisation in an organisation will be key for efficient operations.
- Fully digitized, and streamlined IoT infrastructure, agile operations, and IT capabilities lead to greater supply chain visibility. IoT refers to a system of embedded physical devices connected to a computing environment with the help of software, which results in automatic data appending when sensors get triggered, IoT uses the internet to communicate information to stakeholders.

6. ***Ubiquitous Access of the Data\Information With a Provided Authentication***

- As mentioned above, agile supply chains produce a lot of data about physical & digital assets. Also, it produces valuable insights extracted from captured data in near real-time.
- After collecting a large amount of data, the next concern will be how to maintain this huge data and make it available for customers without latency. Nowadays computing power is cheaper and powerful, we can leverage cloud and in-house servers to store the data and increase the availability of data to customers & stakeholders.

7. ***Quantum Computing and Edge Computing***

- Quantum & Edge computing comes up with high computing performance and deep analysis. Quantum computing is key for specialised predictions and recommendations. Data must be stored on more readily accessible platforms, to utilize industry 4.0 supply chains to their fullest.

8. ***Distributed Ledger Technology (Blockchain)***:

- Blockchain is an open distributed ledger that can record transactions between two parties efficiently and in a verifiable and permanent way (Helo and Hao 2019). It comes up with immutability, decentralized in manner, privacy, increased security, acquaintance, etc.
- Blockchain couples trust with the existing ecosystem to address key challenges, that is achieving visibility, transparency, and traceability, blockchain

is a dynamic decentralized platform, it can be connected to the continuously appending sets of data (Leng et al. 2020).
- Blockchain technology has near real-time settlement of transactions, availability with reliability, immutability as key characteristics.
- Now a days, most of the industries are actively exploring blockchain technology to transform their existing practices to make them efficient and risk free to certain extent. Blockchain technology is opted fit for supply chain digitisation, it has applications in healthcare, food industry, etc. The Unique selling point (USP) of blockchain technology is to integrate the trust into existing ecosystem (Li et al. 2019).
- Use cases of blockchain in supply chains include building coordination between suppliers, digitising logistics, supply chain and logistics visibility, coupling transparency and trust into the existing system, recalling defect-driven products, transparent way of dealing with warranty claims, disintermediation, etc.

11.4 Conclusion

As experts say, "Never let a good crisis go waste" and these times can be a good time to break the silos and see how your firm can transform better to achieve new heights. Though currently, we are living in uncertain and difficult times, the key stakeholders should focus on strategies, frameworks and technologies with which they excel in the near future; nobody knows when this pandemic is going to end, but it is going to come with new challenges and opportunities that might make or break a company. In this study we tried to concentrate on areas dealing with the supply chain VUCAness, new challenges for SC due to the on-going pandemic, long term impacts on SC's due to COVID-19 pandemic, key areas for improvement, achieving maximum possible efficiency in a new normal environment for supply chains, SC practitioners responsibilities to ensure smooth and frictionless operations, adopting digitalisation and emerging technologies for benefit of SC activities, dealing with disruptions in the near future, strategies that can improve the efficiency of SC activities post COVID-19. Here, we discussed the potential challenges for supply chains in the post-COVID era, and discussed what the supply chain practitioners should focus on. The essence of risk management and VUCA has been discussed along with few supply chain networks as examples. Finally, discussed industry 4.0 supply chains and the key steps in achieving digitized supply chains.

References

Attaran, M., & Gunasekaran, A. (2019). Blockchain-enabled technology: The emerging technology set to reshape and decentralise many industries. *International Journal of Applied Decision Sciences, 12*(4), 424–444. https://doi.org/10.1504/IJADS.2019.102642.

Bai, C., & Sarkis, J. (2020). A supply chain transparency and sustainability technology appraisal model for blockchain technology. *International Journal of Production Research, 58*(7), 2142–2162. https://doi.org/10.1080/00207543.2019.1708989.

Bennett, N., & Lemoine, G. J. (2014). What VUCA Really Means for You. Harvard Business Review (JAN-FEB).

Biswas, B., & Gupta, R. (2019). Analysis of barriers to implement blockchain in industry and service sectors. *Computers & Industrial Engineering, 136,* 225–241.

Black swan and perfect stroms. (2012). https://news.stanford.edu/news/2012/november/black-swan-risk-111612.html.

Bürer, M. J., de Lapparent, M., Pallotta, V., Capezzali, M., & Carpita, M. (2019). Use cases for blockchain in the energy industry opportunities of emerging business models and related risks. *Computers & Industrial Engineering, 137,* 106002.

Dolgui, A., Ivanov, D., Potryasaev, S., Sokolov, B., Ivanova, M., & Werner, F. (2020). Blockchainoriented dynamic modelling of smart contract design and execution in the supply chain. *International Journal of Production Research, 58*(7).

Fu, Y., & Zhu, J. (2019). Operation mechanisms for intelligent logistics system: A blockchain perspective. *IEEE Access, 7,* 144202–144213. https://doi.org/10.1109/ACCESS.2019.2945078.

Global COVID-19 information. (2020). https://www.worldometers.info/coronavirus/.

Handbook on eVIN application for vaccine and cold chain manager india. (2016). http://www.cips.org.in/documents/VC/2016/Oct/25/ElectronicVaccineIntelligenceNetwork.pdf.

Helo, P., & Hao, Y. (2019). Blockchains in operations and supply chains: A model and reference implementation. *Computers & Industrial Engineering, 136,* 242–251.

Immunization supply chain and logistics: A neglected but essential system for national immunization programmes. (2020). https://www.who.int/immunization/documents/WHO_IVB_14.05/en/.

Improving cold chain systems: Challenges and solutions. (2020). https://www.sciencedirect.com/science/article/pii/S0264410X16307307.

Leng, J., Yan, D., Liu, Q., Xu, K., Zhao, J. L., Shi, R., et al. (2020). ManuChain: Combining permissioned blockchain with a holistic optimization model as bi-level intelligence for smart manufacturing. *IEEE Transactions on Systems, Man, and Cybernetics: Systems, 50*(1), 182–192. https://doi.org/10.1109/TSMC.2019.2930418.

Li, J., Greenwood, D., & Kassem, M. (2019). Blockchain in the built environment and construction industry: A systematic review, conceptual models and practical use cases. *Automation in Construction, 102,* 288–307.

Paté-Cornell, E. (2012). On "Black Swans" and "Perfect Storms": Risk analysis and management when statistics are not enough. https://doi.org/10.1111/j.1539-6924.2011.01787.x.

Sharma, R., Kamble, S. S., Gunasekaran, A., Kumar, V., & Kumar, A. (2020). A systematic literature review on machine learning applications for sustainable agriculture supply chain performance. *Computers and Operations Research.* (Advance copy).

Who is preparing the next generation of immunization supply chain professionals?. (2020). https://www.ncbi.nlm.nih.gov/pubmed/28364936.